FIRE SAFETY

An employer's guide

London: The Stationery Office

HSE BOOKS

Published with the permission of the Home Office on behalf of the Controller of Her Majesty's Stationery Office.

© Crown copyright 1999
Copyright in the typographical arrangement and design is vested in the Crown.
Application for reproduction should be made in writing to:
Copyright Unit, Her Majesty's Stationery Office
St Clements House, 2-16 Colegate,
Norwich NR3 1BQ

First published 1999

ISBN 0 11 341229 0

All rights reserved. No part of this publication may be reproduced, stored in a retrieval system, or transmitted in any form or by any means (electronic, mechanical, photocopying, recording, or otherwise) without the permission of the copyright owner.

This guidance is issued by the Home Office, the Scottish Executive, the Department of the Environment (Northern Ireland) and the Health and Safety Executive. Following the guidance is not compulsory and you are free to take other action. But if you do follow the guidance you will normally be doing enough to comply with the law.

Acknowledgement

The photographs in this guide are reproduced with the kind permission of London Fire and Civil Defence Authority.

CONTENTS

WHO IS THIS GUIDE FOR? 1
How to use this guide 1

INTRODUCTION 2
An introduction to the law 2

PART 1 - RISK ASSESSMENT 5
What is a risk assessment? 5
How do you do a fire risk assessment? 5
After you have completed your risk assessment 6

PART 2 - THE FIVE STEPS OF FIRE RISK ASSESSMENT 8
Step 1 - Identify fire hazards 8
The fire triangle 8
Identifying sources of ignition 9
Identifying sources of fuel 9
Identifying sources of oxygen 10

Step 2 - Decide who could be harmed 10

Step 3 - Evaluate the risks and decide whether existing precautions are adequate or if more needs to be done 11
How fire spreads through the workplace 11
Reducing sources of ignition 12
Minimising the potential fuel for a fire 13
Reducing sources of oxygen 14
Fire detection and fire warning 14
Means of escape 15
Means of fighting fire 16
Maintenance and testing 17
Fire procedures and training 19
Providing for disabled people 19
Other practical measures 19

Step 4 - Record your findings and actions 21
Your emergency plan 23
Information and instructions for employees 24
Training employees 26

Step 5 - Review and revise 27

PART 3 – FURTHER GUIDANCE ON FIRE PRECAUTIONS 28
Reducing fire risks through good management 28
Maintenance of plant and equipment 28
Storage and use of flammable materials 28
Flammable liquids 29
Work processes involving heat 32
Electrical equipment 34
Heating appliances 35

Smoking and the provision of ashtrays   35
Building and maintenance work   35
Flammable rubbish and waste   36
Reducing the risk of arson   37

Fire detection and warning   38

Means of escape in case of fire   41
Technical terms relating to means of escape   42
Arrangements for evacuating the workplace   43
Assessing means of escape   44
General principles for escape routes   46
Evacuation times and length of escape routes   46
Number and width of exits   48
Inner rooms   48
Corridors   49
Stairways   50
Means of escape for use by staff   52
Reducing the spread of fire, heat and smoke   52
Exhibitions and displays   52
Defining the escape route   53
Items prohibited on an escape route   53
Escape doors   53
Fire doors   54
Emergency escape and fire exit signs   54
Lighting of escape routes   55
Smoke control systems for the safety of people   57
Buildings under refurbishment   57

Fire-fighting equipment   58
Portable fire extinguishers   58
Hose reels and fire blankets   62
Fixed fire-extinguishing systems   62

Installation, maintenance and testing of fire precautions and equipment   64

Historic and listed buildings   68

Disabled people   69

ANNEXES   73
Annex A – Other legislation that may apply to your workplace   73
Annex B – Enforcement of the Fire Regulations   77

REFERENCES   78

FURTHER INFORMATION   83

WHO IS THIS GUIDE FOR?

This guide tells you, the employer, what you have to do to comply with the law relating to fire issues. It also tells you how to carry out your fire risk assessment and identify the safeguards which you should have in your workplace.

Although written for employers, the guide will also be useful if you are self-employed or are in control of workplaces to which people you do not employ, and members of the public, have access. The information will also provide a useful source of reference for:

- employees;
- employee-elected representatives;
- trade union-appointed health and safety representatives; and
- all other people who have a role in ensuring fire safety in the workplace.

The guide does not cover construction sites; ships and other means of transport; mines; offshore installations; and any workplace in fields, woods or agricultural land away from main buildings.

HOW TO USE THIS GUIDE

This guide is divided into sections which are intended to help you understand what you need to do to comply with fire safety law. It also provides further advice on fire precautions which you can use if you wish or need to.

Part 1 explains what fire risk assessment is and how you might go about it. Fire risk assessment should be the foundation for all the fire precautions in your workplace.

Part 2 is about fire risk assessment and leads you step by step through the assessment process.

Part 3 provides further guidance on fire precautions. The information is provided for employers and others to dip into during their fire risk assessment or when they are reviewing precautions.

INTRODUCTION

This guidance gives advice on how to avoid fires in the workplace and how to ensure people's safety if a fire does start. Why should you read it? Because:

- Fire kills. In 1997, UK fire brigades attended over 36 000 fires in workplaces. These fires killed 30 people and injured over 2600.
- Fire costs money. The costs of a serious fire can be high and afterwards many businesses do not reopen. You can get advice about minimising fire losses from your insurer, or the Fire Protection Association (see Further information section).

This guidance explains the basic requirements of and can help you comply with:

- the Fire Precautions (Workplace) Regulations 1997 (as amended) (in Northern Ireland the Fire Precautions (Workplace) (Northern Ireland) Regulations*) - these are referred to as the Fire Regulations in the rest of this guide; and
- the Management of Health and Safety at Work Regulations 1992 (as amended).

The advice in this guide is applicable to most workplaces. However, not all the precautions will be relevant in all circumstances, nor is the guidance intended to be sufficiently comprehensive to cover those workplaces where very large quantities of flammable or explosive materials are handled. Guidance on these is listed in the References section. If you need any further information, contact your relevant enforcing authority.

AN INTRODUCTION TO THE LAW

The term 'fire precautions' includes matters which are the subject of legal requirements under specific fire precautions legislation. These include the Fire Regulations and the Fire Precautions Act 1971 (in Northern Ireland the Fire Services (Northern Ireland) Order 1984 (as amended)) and, more generally, under health and safety legislation including the Health and Safety at Work etc Act 1974 and regulations made under that Act.

The Fire Regulations and the Fire Precautions Act 1971 (and Northern Ireland equivalents) are the responsibility of the Home Departments and are enforced by the fire authorities. However, for Crown-occupied and Crown-owned premises, enforcement is carried out by the Fire Service Inspectorates of the Home Departments (in Northern Ireland by the Department of Economic Development).

Fire precautions legislation deals with general fire precautions. These include:

- means of detection and giving warning in case of fire;
- the provision of means of escape;
- means of fighting fire; and
- the training of staff in fire safety.

The Fire Regulations also include a requirement to undertake an assessment of the fire risks. (In this guide, the term 'fire risk' collectively describes both the risk of fire occurring and the risk to people in the event of fire.)

* The Northern Ireland Regulations are due to be made in 1999.

Introduction

The Health and Safety at Work etc Act 1974 and regulations made under it cover the provision of process fire precautions which are intended to prevent the outbreak of a fire or minimise the consequences should one occur. Matters falling within the scope of the Act include the storage of flammable materials, the control of flammable vapours, standards of housekeeping, safe systems of work, the control of sources of ignition and the provision of appropriate training. These precautions are enforced by inspectors from the Health and Safety Executive (HSE) or the local authority.

Between them, the Fire Regulations and the Management of Health and Safety at Work Regulations 1992 (as amended) require you to:

- carry out a fire risk assessment of your workplace (you must consider all your employees and all other people who may be affected by a fire in the workplace and you are required to make adequate provision for any disabled people with special needs who use or may be present at your premises);
- identify the significant findings of the risk assessment and the details of anyone who might be especially at risk in case of fire (these must be recorded if you employ more than five people);
- provide and maintain such fire precautions as are necessary to safeguard those who use your workplace; and
- provide information, instruction and training to your employees about the fire precautions in your workplace.

The risk assessment will help you decide the nature and extent of the general and process fire precautions which you need to provide.

There are six other legal duties you need to know:

- Where it is necessary to safeguard the safety of your employees, you must nominate people to undertake any special roles which are required under your emergency plan (you can nominate yourself for this purpose).
- You must consult your employees (or their elected representatives or appointed trade union safety representatives) about the nomination of people to carry out particular roles in connection with fire safety and about proposals for improving the fire precautions.
- You must inform other employers who also have workplaces in the building of any significant risks you found which might affect the safety of their employees - and co-operate with them about the measures proposed to reduce/control those risks.
- If you are not an employer but have any control of premises which contain more than one workplace, you are also responsible for ensuring that the requirements of the Fire Regulations are complied with in those parts you have control over.
- You must establish a suitable means of contacting the emergency services, and ensure that they can be called easily.
- The law requires your employees to co-operate with you to ensure the workplace is safe from fire and its effects, and not to do anything which will place themselves or other people at risk.

Introduction

In some workplaces you may need to apply for a fire certificate, a licence, or other form of approval before using the workplace. You can find more information about the legislation which might apply to your workplace in Annex A, and enforcement arrangements for the Fire Regulations in Annex B.

PART 1 - RISK ASSESSMENT

WHAT IS A RISK ASSESSMENT?

It is an organised look at what, in your work activities and workplace, could cause harm to people. This will allow you to weigh up whether you have taken enough precautions or should do more to avoid harm. The important things you need to decide are whether a hazard is significant and whether you have covered it by satisfactory precautions so that the risk is acceptably low.

What do the terms 'hazard' and 'risk' mean?

- A hazard is something that has the potential to cause harm.
- A risk is the chance, high or low, of that harm occurring.

Before you start your risk assessment

Check whether any of the fire safety arrangements in your workplace have previously been approved under other fire safety, licensing or building legislation. If this is the case, an assessment of the fire precautions needed under that legislation will have been made at the time by, or in consultation with, the fire authority or the building control authority.

Regardless of any previous approval, you still need to carry out a fire risk assessment. However, if the previous approval covered all the matters required by the Fire Regulations, and conditions have remained unchanged, eg numbers of people present, work activity etc, then your fire risk assessment may well show that few, if any, additional precautions are needed.

Your risk assessment may identify additional matters which need addressing if the previous approval was given according to an out-of-date standard of fire precautions, or the approval was under legislation which does not cover all the requirements of the Fire Regulations. If you are not sure, your local fire authority will be able to advise you.

HOW DO YOU DO A FIRE RISK ASSESSMENT?

A fire risk assessment will help you determine the chances of a fire occurring and the dangers from fire that your workplace poses for the people who use it. The assessment method suggested shares the same approach as that used in general health and safety legislation and can be carried out either as part of a more general risk assessment or as a separate exercise.

Before attempting to start an assessment take time to prepare. Read through the rest of Parts 1 and 2 of this guide and plan how you will go about your assessment.

A risk assessment is not a theoretical exercise. However, much work can be done on paper from the knowledge you, your employees or their representatives have of the workplace. A tour of the workplace will be needed to confirm, amend or add detail to your initial views.

Part 1 - Risk assessment

For fire risk assessments there are five steps that you need to take:

Step 1 Identify potential fire hazards in the workplace.
Step 2 Decide who (eg employees, visitors) might be in danger, in the event of a fire, in the workplace or while trying to escape from it, and note their location.
Step 3 Evaluate the risks arising from the hazards and decide whether your existing fire precautions are adequate or whether more should be done to get rid of the hazard or to control the risks (eg by improving the fire precautions).
Step 4 Record your findings and details of the action you took as a result. Tell your employees about your findings.
Step 5 Keep the assessment under review and revise it when necessary.

Nobody knows as much about your business as you and the people who work with and for you. Try to use your own knowledge and experience and that of your colleagues and staff. Talk to your employees and listen to their concerns. The safety representative (if there is one) and your employees will have a valuable contribution to make. They can help you identify key issues and may already have practical suggestions for improvements.

Proper planning of your assessment, and any changes necessary because of it, includes consulting the workforce and their representatives. This can help ensure that any changes are introduced more easily and accepted more readily. However, remember that risk assessment is essentially a matter of applying informed common sense. You need to identify what could reasonably be expected to cause danger. Ignore the trivial and concentrate on significant hazards.

It is important that you carry out your fire risk assessment in a practical and systematic way. It must take the whole of the workplace into account, including outdoor locations and any rooms and areas which are rarely used. If your workplace is small you may be able to assess the workplace as a whole. In larger buildings, you will often find it helpful to divide the workplace into rooms or a series of assessment areas using natural boundaries, eg process areas, offices, stores, workshops as well as corridors, stairways and external routes.

If your workplace is in a building shared with other employers, you and all the other occupiers and any other person who has control of any other part of the workplace will need to discuss your risk assessments. This will help to ensure that any areas of higher risk, and the need for any extra precautions, are identified.

AFTER YOU HAVE COMPLETED YOUR RISK ASSESSMENT

If you know, or think, that your workplace is subject to a fire certification or licensing regime, as well as the Fire Regulations, you will need to check that any changes you propose as a result of your risk assessment will not conflict with this other regime. You need to do this before making any changes. In these cases you should consult the local fire

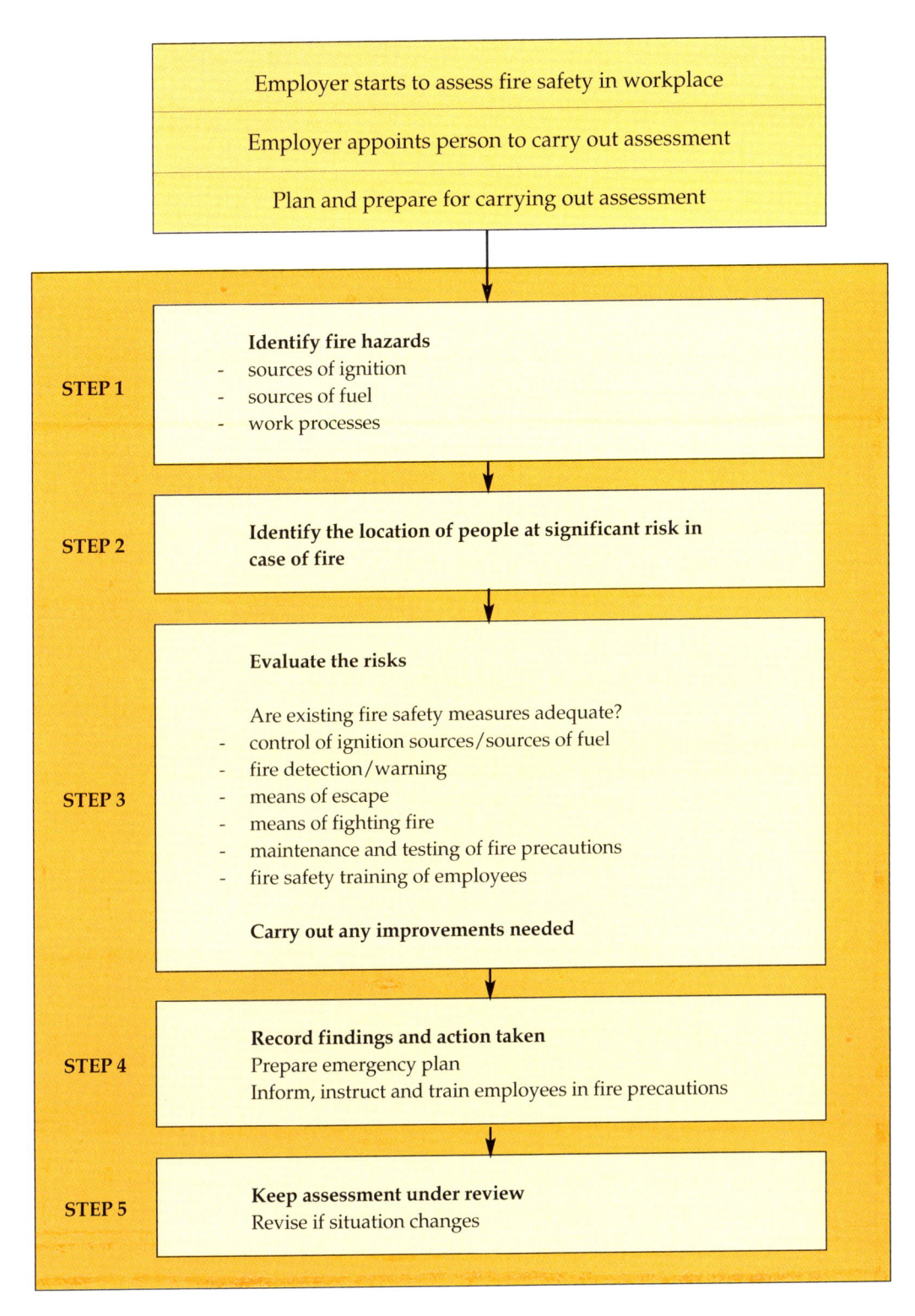

Part 1 - Risk assessment

```
┌─────────────────────────────────────────────────┐
│  Employer starts to assess fire safety in workplace │
├─────────────────────────────────────────────────┤
│  Employer appoints person to carry out assessment   │
├─────────────────────────────────────────────────┤
│  Plan and prepare for carrying out assessment       │
└─────────────────────────────────────────────────┘
```

STEP 1 — **Identify fire hazards**
- sources of ignition
- sources of fuel
- work processes

STEP 2 — **Identify the location of people at significant risk in case of fire**

STEP 3 — **Evaluate the risks**

Are existing fire safety measures adequate?
- control of ignition sources/sources of fuel
- fire detection/warning
- means of escape
- means of fighting fire
- maintenance and testing of fire precautions
- fire safety training of employees

Carry out any improvements needed

STEP 4 — **Record findings and action taken**
Prepare emergency plan
Inform, instruct and train employees in fire precautions

STEP 5 — **Keep assessment under review**
Revise if situation changes

Flowchart of an action plan for risk assessment

authority. They will consider your proposals and advise you if they are acceptable. They will also advise you if any other legislative approvals have to be obtained. For instance, if you propose structural alteration or material change of use (in Scotland, any changes) to a building, approval under relevant building legislation will be required.

PART 2 - THE FIVE STEPS OF FIRE RISK ASSESSMENT

STEP 1 - IDENTIFY FIRE HAZARDS

The fire triangle

For a fire to start, three things are needed:

- a source of ignition;
- fuel; and
- oxygen.

If any one of these is missing, a fire cannot start. Taking steps to avoid the three coming together will therefore reduce the chances of a fire occurring.

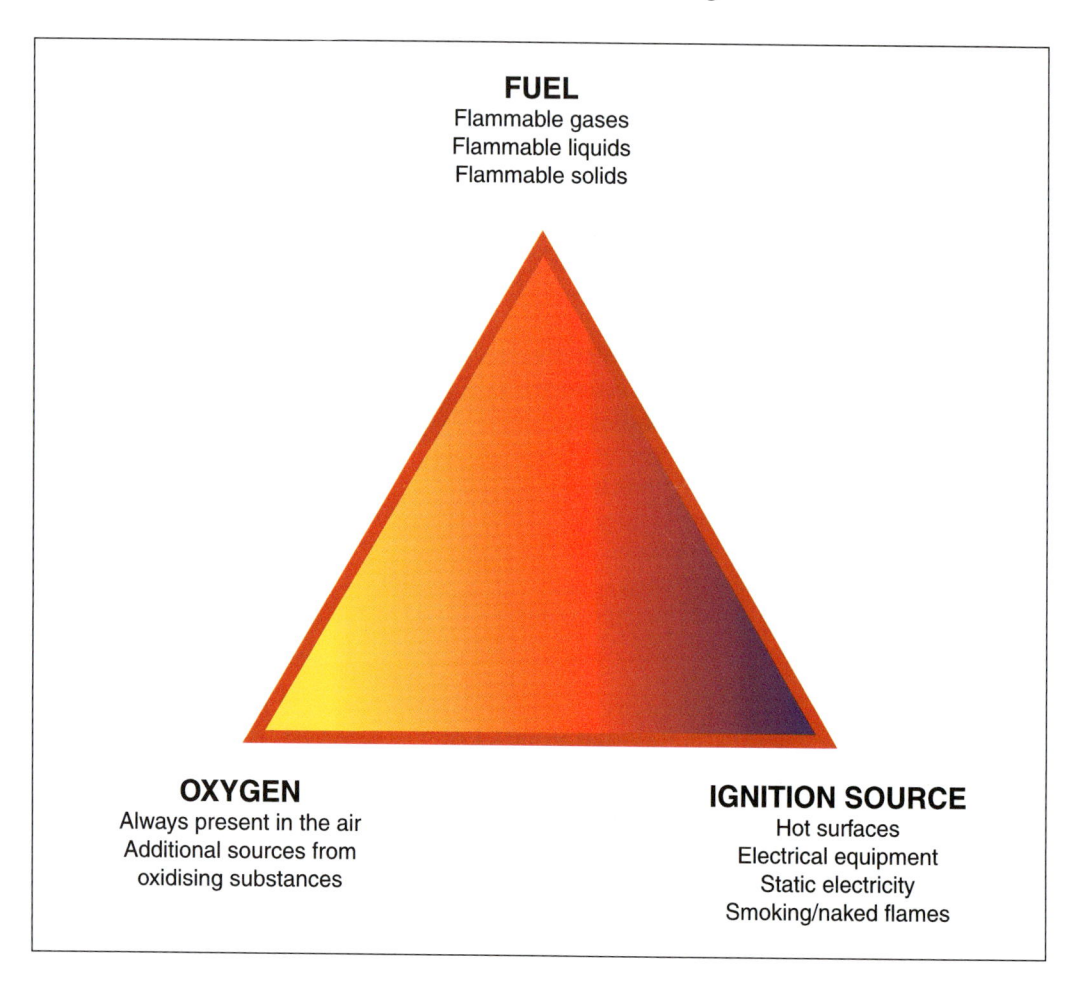

FUEL
Flammable gases
Flammable liquids
Flammable solids

OXYGEN
Always present in the air
Additional sources from oxidising substances

IGNITION SOURCE
Hot surfaces
Electrical equipment
Static electricity
Smoking/naked flames

Once a fire starts it can grow very quickly and spread from one source of fuel to another. As it grows, the amount of heat it gives off will increase and this can cause other fuels to self-ignite.

The following paragraphs advise on how to identify potential ignition sources, the materials that might fuel a fire and the oxygen supplies which will help it to burn.

Part 2 - The five steps of fire risk assessment

Identifying sources of ignition

You can identify the potential ignition sources in your workplace by looking for possible sources of heat which could get hot enough to ignite the material in the workplace. These sources of heat could include:

- smokers' materials, eg cigarettes and matches;
- naked flames;
- electrical, gas or oil-fired heaters (fixed or portable);
- hot processes (such as welding or grinding work);
- cooking;
- engines or boilers;
- machinery;
- faulty or misused electrical equipment;
- lighting equipment, eg halogen lamps;
- hot surfaces and obstruction of equipment ventilation, eg office equipment;
- friction, eg from loose bearings or drive belts;
- static electricity;
- metal impact (such as metal tools striking each other); and
- arson.

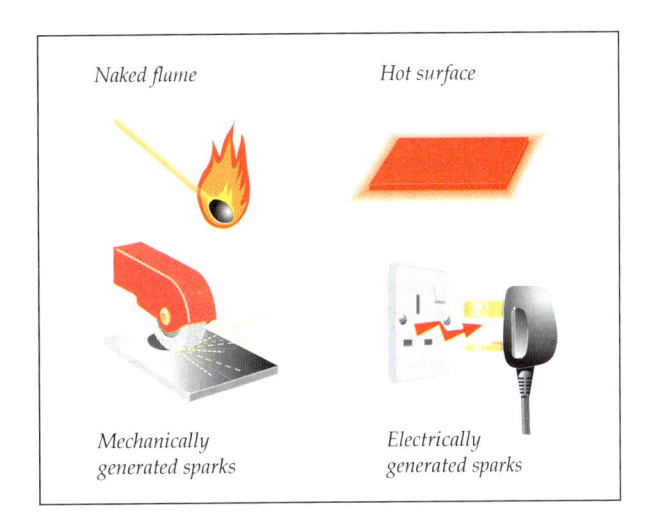

Naked flame   *Hot surface*

Mechanically generated sparks   *Electrically generated sparks*

Indications of 'near misses', such as scorch marks on furniture or fittings, discoloured or charred electrical plugs and sockets, cigarette burns etc, can help you identify hazards which you may not otherwise notice.

Identifying sources of fuel

Anything that burns is fuel for a fire. So you need to look for the things that will burn reasonably easily and are in sufficient quantity to provide fuel for a fire or cause it to spread to another fuel source. Some of the most common 'fuels' found in workplaces are:

- flammable liquid based products such as paints, varnish, thinners and adhesives;
- flammable liquids and solvents such as petrol, white spirit, methylated spirit and paraffin;
- flammable chemicals;
- wood;
- paper and card;
- plastics, rubber and foam such as polystyrene and polyurethane, eg the foam used in upholstered furniture;
- flammable gases such as liquefied petroleum gas (LPG) and acetylene;

9

Part 2 - The five steps of fire risk assessment

- furniture, including fixtures and fittings;
- textiles;
- loose packaging material; and
- waste materials, in particular finely divided materials such as wood shavings, offcuts, dust, paper and textiles.

You should also consider the construction of your workplace and how this might contribute to the spread of fire. Does the internal construction include large areas of:

- hardboard, chipboard, blockboard walls or ceilings; or
- synthetic ceiling or wall coverings, such as polystyrene tiles?

If these are present, and you are uncertain of the danger they might pose, you should seek advice from your local fire authority or other experts on what precautions you need to take to reduce the risk to people in the event of fire.

Identifying sources of oxygen

The main source of oxygen for a fire is in the air around us. In an enclosed building this is provided by the ventilation system in use. This generally falls into one of two categories: natural airflow through doors, windows and other openings; or mechanical air conditioning systems and air handling systems. In many buildings there will be a combination of systems, which will be capable of introducing/extracting air to and from the building.

Additional sources of oxygen can sometimes be found in materials used or stored in a workplace such as:

- some chemicals (oxidising materials), which can provide a fire with additional oxygen and so assist it to burn. These chemicals should be identified on their container by the manufacturer or supplier who can advise as to their safe use and storage; or
- oxygen supplies from cylinder storage and piped systems, eg oxygen used in welding processes or for health care purposes.

STEP 2 - DECIDE WHO COULD BE HARMED

If there is a fire, the main priority is to ensure that everyone reaches a place of safety quickly. Putting the fire out is secondary to this because the greatest danger from fire in a workplace is the spread of the fire, heat and smoke through it. If a workplace does not have adequate means of detecting and giving warning or means of escape, a fire can trap people or they may be overcome by the heat and smoke before they can evacuate.

As part of your assessment, you need to identify who may be at risk if there is a fire, how they will be warned and how they will escape. To do this you need to identify where you have people working, whether at permanent workstations or occasional ones, and to consider who else might be at risk, such as customers, visiting contractors etc, and where these people are likely to be found.

Part 2 - The five steps of fire risk assessment

STEP 3 - EVALUATE THE RISKS AND DECIDE WHETHER EXISTING PRECAUTIONS ARE ADEQUATE OR IF MORE NEEDS TO BE DONE

Steps 1 and 2 will have helped you to identify what the hazards are and who may be at risk because of them. You now need to evaluate the risk and decide whether you have done enough to reduce this or need to do more by considering:

- the chance of a fire occurring and whether you can reduce the sources of ignition/minimise the potential fuel for a fire;
- the fire precautions you have in place and whether they are sufficient for the remaining risks and will ensure everyone is warned in case of a fire; and
- the means people can use to make their escape safely (or put the fire out if it is safe for them to do so).

How fire spreads through the workplace

To be able to assess whether people will be at risk in the event of a fire it helps to have an appreciation of the risks posed as it develops. Most people will be familiar with a fire outdoors, such as a bonfire, which people can move back from as it grows. If the wind is blowing the smoke towards them, they can move right away from the fire to a place of safety because they have a choice of escape routes not affected by heat or smoke.

Fires in enclosed spaces, such as buildings, behave differently to fires in the open air. The smoke rising from the fire gets trapped by the ceiling and then spreads in all directions to form an ever-deepening layer over the entire room or space. During this process, the smoke will pass through any holes or gaps in the walls, ceiling or floor and eventually into other parts of the workplace. The heat from the fire also gets trapped in the building, greatly increasing the temperature.

Spread of smoke through a building

11

Part 2 - The five steps of fire risk assessment

There is an added danger to people due to the toxic gases in the smoke produced by a fire. People are therefore at a greater risk from a fire indoors than one outdoors. It is essential that the means of escape and other fire precautions are adequate to ensure that everyone can make their escape to a place of safety before the fire and its effects can trap them in the building.

It is essential that the start of any fire is detected as quickly as possible and certainly before it can make the means of escape unusable. In some circumstances, particularly where people are located away from the origin of the fire and there is a reasonable possibility that it could spread, this could mean that the fire may need to be detected within two minutes of it starting. This is so that people have enough time to escape safely. Where quantities of highly flammable liquids or gases are involved, it may be necessary to detect any fire in seconds rather than minutes. Once a fire has been detected, the people in your workplace should be signalled to evacuate the building. (There is more information in 'Fire detection and warning' in Part 3.)

You need to give particular attention to:

- any areas, particularly unoccupied ones, where there could be a delay in detecting the start of a fire;
- any areas where the warning may go unnoticed; and
- people who may be unable to react quickly.

The following paragraphs indicate some possible actions for reducing fire hazards and will provide some useful guidance to help you assess the adequacy of your own fire precautions. But remember that the fire risk assessment is an ongoing process and is a means and not an end. If your assessment shows that you need to do more to control risk, then you should do it.

Reducing sources of ignition

You can reduce the hazards caused by potential sources of heat by:

- removing unnecessary sources of heat from the workplace or replacing them with safer alternatives, ensuring that heat-producing equipment is used in accordance with the manufacturer's instructions and is properly maintained;
- installing machinery and equipment which has been designed to minimise the risk of fire and explosions;
- replacing naked flame and radiant heaters with fixed convector heaters or a central heating system;
- ensuring that all electrical fuses and circuit breakers etc are of the correct rating and suitable for the purpose;
- ensuring that sources of heat do not arise from faulty or overloaded electrical or mechanical equipment (such as overheating bearings);
- keeping ducts and flues clean;

Part 2 - The five steps of fire risk assessment

- where appropriate, operating a permit to work system for maintenance workers and contractors who carry out 'hot work' involving processes such as welding or flame cutting;
- operating a safe smoking policy in designated smoking areas and prohibiting smoking elsewhere;
- enforcing the prohibition of matches and lighters and other naked flames in high-fire-risk areas;
- ensuring that all equipment that could provide a source of ignition, even when not in use, is left in a safe condition;
- making sure that any smouldering material (including smokers' material) is properly extinguished before leaving the workplace; and
- taking precautions to avoid the risk of arson.

Further information about reducing sources of ignition is given in Part 3 (see 'Reducing fire risks through good management').

Minimising the potential fuel for a fire

There are various ways you can reduce the risks caused by materials and substances which burn. These include:

- removing flammable materials and substances, or reducing them to the minimum required for the operation of the business;
- replacing materials and substances with less flammable alternatives;
- ensuring flammable materials, liquids (and vapours) and gases are handled, transported, stored and used properly;
- ensuring adequate separation distances between flammable materials;
- storing highly flammable substances in fire-resisting stores and, where necessary, keeping a minimum quantity in fire-resisting cabinets in the workroom;

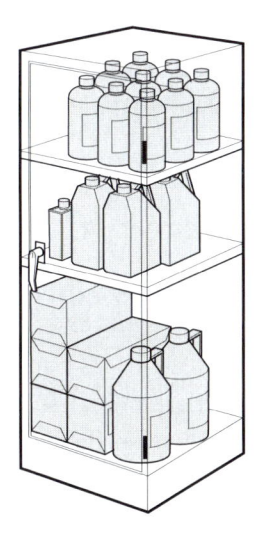

Safe storage of small quantities of highly flammable substances in fire-resisting cabinets

13

Part 2 - The five steps of fire risk assessment

- removing, covering or treating large areas of flammable wall and ceiling linings to reduce the rate of flame spread across the surface;
- replacing or repairing furniture with damaged upholstery where the foam filling is exposed;
- ensuring that flammable waste materials and rubbish are not allowed to build up and are carefully stored until properly disposed of;
- taking action to avoid storage areas being vulnerable to arson or vandalism;
- ensuring good housekeeping; and
- improving the fire-resistance of the construction of the workplace.

Reducing sources of oxygen

You can reduce the potential source of oxygen supply to a fire by:

- closing all doors, windows and other openings not required for ventilation, particularly out of working hours;
- shutting down ventilation systems which are not essential to the function of the workplace;
- not storing oxidising materials near or with any heat source or flammable materials; and
- controlling the use and storage of oxygen cylinders, ensuring that they are not leaking, are not used to 'sweeten' the atmosphere, and that where they are located is adequately ventilated.

Fire detection and fire warning

You need to have an effective means of detecting any outbreak of fire and for warning people in your workplace quickly enough so that they can escape to a safe place before the fire is likely to make escape routes unusable.

In small workplaces where a fire is unlikely to cut off the means of escape, eg open-air areas and single-storey buildings where all exits are visible and the distances to be travelled are small, it is likely that any fire will be quickly detected by the people present and a shout of 'Fire!' may be all that is needed.

In larger premises, particularly multi-storey premises, an electrical fire warning system with manually operated call points is likely to be the minimum needed. In unoccupied areas, where a fire could start and develop to the extent that escape routes may become affected before it is discovered, it is likely that a form of automatic fire detection will also be necessary.

In some cases where a fire certificate or licence is in force, the existing arrangements may be satisfactory (see 'Before you start your risk assessment' in Part 1).

Part 2 - The five steps of fire risk assessment

Checklist

- Can the existing means of detection discover a fire quickly enough to raise an alarm in time for all the occupants to escape to a safe place?
- Can the means for giving warning be clearly heard and understood throughout the whole premises when initiated from any single point?
- If the fire detection and warning system is electrically powered, does it have a back-up power supply?
- Have you told your employees about your fire warning system, will they know how to operate it and respond to it?
- Are there instructions for your employees on how to operate the fire warning system and what action they should take on hearing a warning?
- Have you included the fire detection and fire warning arrangements in your emergency plan? (See Step 4.)

If you are not sure about your current arrangements you should refer to the further guidance in 'Fire detection and warning' in Part 3.

Means of escape

Once a fire has been detected and a warning given, everyone in your workplace should be able to evacuate without being placed at undue risk.

In buildings, most deaths from fire are due to the inhalation of smoke. Also, where smoke is present, people are often unwilling to travel more than a few metres through it to make their escape. It is therefore important to make sure that, in the event of a fire in one part, people in other parts of the building can use escape routes to get out safely without being exposed to the smoke or gases from the fire.

When assessing the means of escape from your workplace, you should ask yourself whether people could escape to a place of safety before being cut off and exposed to risk of injury.

In small, single-storey premises, where travel distances are short, the time taken for people to escape once they are aware of the fire can often be measured in seconds rather than minutes. So it is likely that the normal exits will be sufficient in an emergency. In larger premises, where travel distances are greater and where it is possible for a single route to be affected, an alternative means of escape will normally be necessary.

In most cases where the means of escape has recently been approved under building legislation, a fire certificate or a licence, the existing arrangements will probably be satisfactory. If your risk assessment suggests that change may be necessary, you should check what you propose with the fire authority.

Part 2 - The five steps of fire risk assessment

Checklist

- How long will it take for all the occupants to escape to a place of safety once a fire has been detected?
- Is that a reasonable length of time or will it take too long?
- Are there enough exits and are they in the right place?
- Are the type and size of exits suitable and sufficient for the number of people likely to need to use them (eg wide enough for wheelchair users)?
- In the event of fire, could all available exits be affected or will at least one route from any part of the premises remain available?
- Are all escape routes easily identifiable, free from any obstructions and adequately illuminated?
- Have you trained your staff in using the means of escape?
- Are there instructions about the means of escape for your employees?
- Have you included your means of escape arrangements in your emergency plan? (See Step 4.)

If you are not sure about your current arrangements, you should refer to the further guidance in 'Means of escape in case of fire' in Part 3.

Keep both sides of emergency exits free from obstruction

Means of fighting fire

You need to have enough fire-fighting equipment in place for your employees to use, without exposing themselves to danger, to extinguish a fire in its early stages. The equipment must be suitable to the risks and appropriate staff will need training and

Part 2 - The five steps of fire risk assessment

instruction in its proper use. In small premises, having one or two portable extinguishers in an obvious location may be all that is required.

In larger or more complex premises, a greater number of portable extinguishers, strategically sited throughout the premises, are likely to be the minimum required. Other means of fighting fire may need to be considered and details of these are given in 'Fire-fighting equipment' in Part 3.

In premises where a fire certificate or a licence is in force, it is likely that the existing arrangements will be satisfactory.

Checklist

- Are the extinguishers suitable for the purpose and of sufficient capacity?
- Are there sufficient extinguishers sited throughout the workplace?
- Are the right types of extinguishers located close to the fire hazards and can users gain access to them without exposing themselves to risk?
- Are the locations of the extinguishers obvious or does their position need indicating?
- Have the people likely to use the fire extinguishers been given adequate instruction and training?
- Have you included use of fire-fighting equipment in your emergency plan? (See Step 4.)

If you are not sure about your current arrangements, see the further guidance in 'Fire-fighting equipment' in Part 3.

Maintenance and testing

You need to keep the fire safety measures and equipment in the workplace in effective working order. This includes all fixtures and fittings such as fire doors, staircases, corridors, fire detection and alarm systems, fire-fighting equipment, notices and emergency lighting. You need to carry out regular checks, periodic servicing and maintenance whatever the size of the workplace. Any defects should be put right as quickly as possible.

You, or an employee you have nominated, can carry out checks and routine maintenance work. However, it is important to ensure the reliability and safe operation of fire-fighting equipment and installed systems such as fire alarms and emergency lighting. This is best done by using a competent person* to carry out periodic servicing and any necessary repairs. A record of the work carried out on such equipment and systems will help to demonstrate compliance with the law.

If your premises are covered by a fire certificate, requirements are likely to have been imposed which cover all the equipment you have. Provided you have fulfilled all the requirements of your fire certificate, it is unlikely that you will need to do any more.

---

* A competent person is someone who has the necessary knowledge, training, experience and abilities to carry out the work.

Part 2 - The five steps of fire risk assessment

Checklist

- Do you regularly check all fire doors and escape routes and associated lighting and signs?
- Do you regularly check all your fire-fighting equipment?
- Do you regularly check your fire detection and alarm equipment?
- Do you regularly check any other equipment provided to help means of escape arrangements in the building?
- Are there instructions for relevant employees about testing of equipment?
- Are those who test and maintain the equipment properly trained to do so?

The following table describes good practice for the routine maintenance and testing of fire safety equipment. All other fixtures and fittings, such as fire doors, staircases, corridors and notices to assist safe escape from the workplace, should be regularly checked. Any defects found should be put right as quickly as possible. You can find further information in 'Installation, maintenance and testing of fire precautions and equipment' in Part 3.

Table 1: Maintenance of fire precautions

Equipment	Period	Action
Fire detection and fire warning systems including self-contained smoke alarms and manually operated devices.	Weekly	Check all systems for state of repair and operation. Repair or replace defective units. Test operation of systems, self-contained alarms and manually operated devices.
	Annually	Full check and test of system by competent service engineer. Clean self-contained smoke alarms and change batteries.
Emergency lighting equipment including self-contained units and torches.	Weekly	Operate torches and replace batteries as required. Repair or replace any defective unit.
	Monthly	Check all systems, units and torches for state of repair and apparent working order.
	Annually	Full check and test of systems and units by competent service engineer. Replace batteries in torches.
Fire-fighting equipment including hose reels.	Weekly	Check all extinguishers including hose reels for correct installation and apparent working order.
	Annually	Full check and test by competent service engineer.

Note: Unless otherwise stated, the above actions can be carried out by the user. Manufacturers may recommend alternative or additional action. Further, more detailed information can be found in the relevant British Standards (see the References section).

Part 2 - The five steps of fire risk assessment

Fire procedures and training

In the event of a fire your employees will need to know what to do. You will need to have adequate pre-planned procedures in place (your emergency plan) and ensure your employees are trained in line with those procedures. The procedures will also need to be regularly tested (see Step 4).

In small premises the procedures required may be relatively simple, but in larger, more complex premises they will need to be more comprehensive. In all cases, the emergency plan will need to take account of all people likely to be in the premises at any time (a shop may have a small number of employees but larger numbers of customers).

If you have a fire certificate there will be certain requirements imposed by the certificate concerning procedures and training. Provided you have complied with these requirements, you may not need to do any more.

Checklist

- Do you have an emergency plan?
- Does the emergency plan take account of all reasonably foreseeable circumstances?
- Are your employees familiar with the plan, trained in its use and involved in testing it?
- Is the emergency plan made available to all who need to be aware of it?
- Are the procedures to be followed clearly indicated throughout the workplace?
- Have you considered all the people likely to be present in your workplace and others with whom you may share the building?

If you are in any doubt about your current arrangements, you should refer to the guidance in Step 4 and the information in 'Arrangements for evacuating the workplace' in Part 3.

Providing for disabled people

You should make adequate provision for people with special needs who may be present in your premises. You need to consider both employees and visitors, and take into account not only people who have restricted mobility but also those who have poor hearing, poor sight or any other form of disability. Further guidance is given in 'Disabled people' in Part 3.

Other practical measures

In addition to basic, physical fire precautions such as means of escape, there are other things that you can do to ensure a quick and safe evacuation of the workplace. These include:

Part 2 - The five steps of fire risk assessment

- ensuring good housekeeping;
- ensuring escape routes are kept clear;
- ensuring suitable security measures to reduce the chance of arson; and
- having regular fire safety checks.

Housekeeping

Do not store anything in escape corridors, stairways or lobbies, even temporarily, which could cause an obstruction and hinder safe escape from the workplace. Ensure there are no flammable materials, including rubbish and waste, in these areas as they could support the spread of fire, making the escape route unsafe. Further advice is given in 'Reducing fire risks through good management' and 'Means of escape in case of fire' in Part 3.

Security: reducing the risk of arson

The risk of arson can be reduced by having good security and by ensuring that any flammable materials stored outside the building do not put the workplace at risk (see 'Reducing the risk of arson' in Part 3).

Part 2 - The five steps of fire risk assessment

Fire safety checks

Although this guidance is about reducing the risks of fire and protecting people against the risk from fire, many of the precautions recommended will also help you protect your workplace and its contents while it is unoccupied. You can do this by ensuring that a full check is carried out to make sure that the workplace is left in a safe condition before it is vacated. This should include checking that:

- all windows and doors are closed, including doors held open by automatic release units;
- electrical equipment not in use is switched off, and where appropriate, unplugged;
- smokers' materials are not left smouldering;
- all naked flames are extinguished or left in a safe condition;
- all flammable rubbish and waste is removed to a safe place;
- all highly flammable materials are safely stored; and
- the workplace is secured against unauthorised entry.

STEP 4 - RECORD YOUR FINDINGS AND ACTIONS

If you employ five or more employees you must record the significant findings of your risk assessment, together with details of any people you identify as being at particular risk. You will probably find it useful (unless your assessment is very simple) to keep a written record of your fire risk assessment as you go round. This will help you plan the actions you need to take in the light of the findings of your risk assessment.

This record might take the form of a simple list, or you could use a plan of the layout of the workplace, or a combination of both of these.

Significant hazards	People/groups of people who are at risk from the hazards	Existing controls and risks which are not adequately controlled	What further action is needed - by when? - by who?

Example of a simple list

Part 2 - The five steps of fire risk assessment

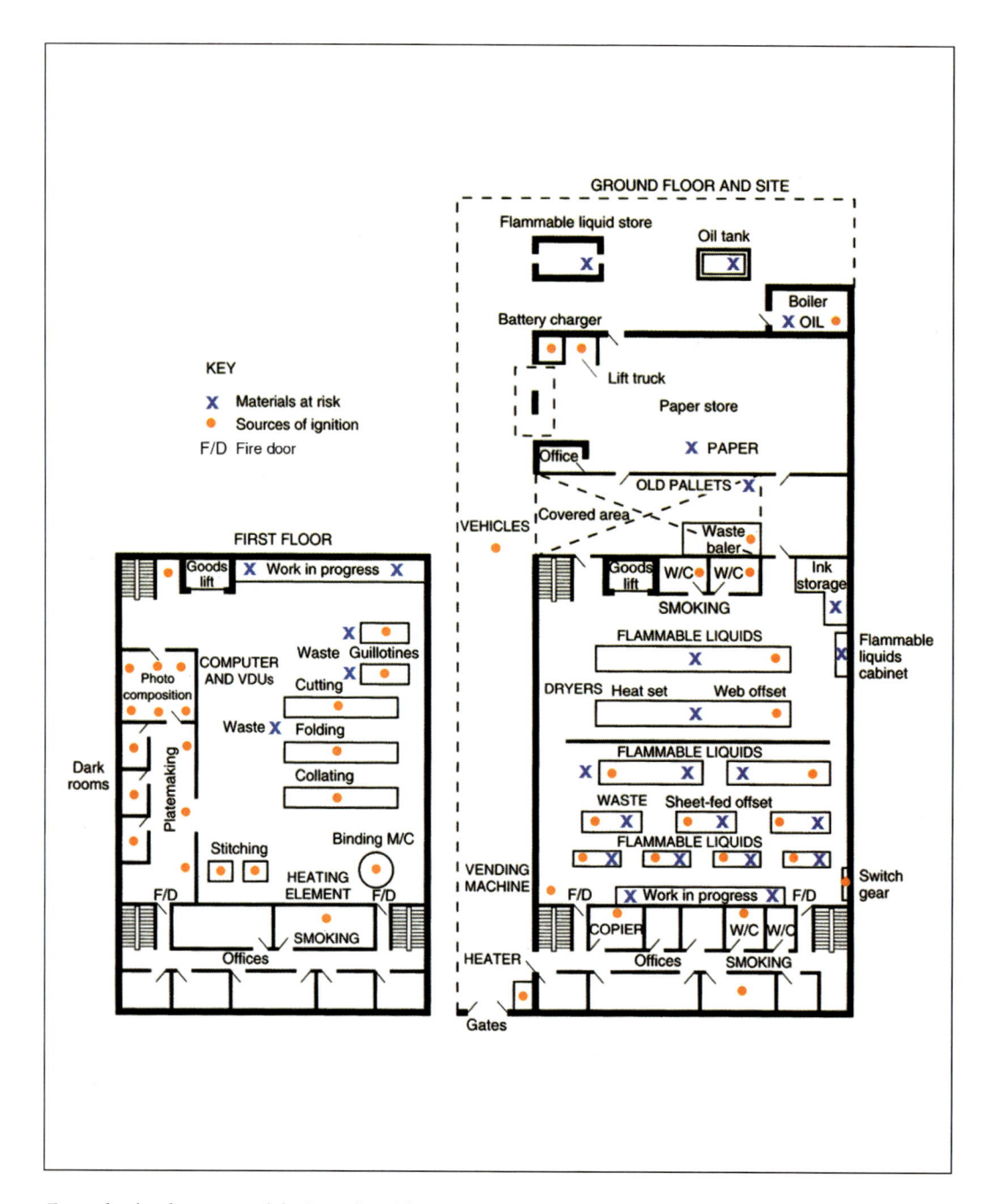

Example of a plan prepared during a fire risk assessment

Having completed your assessment and put your fire precautions in place, it can be useful to record details of maintenance and testing work carried out on them. It is also helpful to record details of the instruction and training you give to employees and when they took place. Although these are not requirements under the Fire Regulations, you may be required to keep such records under other legislation, eg if you have a fire certificate issued under the Fire Precautions Act 1971. Such records can assist you, particularly when reviewing your assessment. They also show the relevant enforcing authorities the actions you have taken to comply with the Fire Regulations and other fire safety legislation.

Part 2 - The five steps of fire risk assessment

The date of the training or drill	
Duration of training	
Fire drill evacuation times	
Name of person giving instruction	
Names of people receiving instructions	
The nature of the instruction or drill	
Any observations/remedial action	

Example of a training record

Your emergency plan

You need to plan the action that your employees and other people in the workplace should take in the event of a fire. If you employ more than five people then you must have a written emergency plan. This emergency plan should be kept in the workplace, be available to your employees and the employees' representatives (where appointed) and form the basis of the training and instruction you provide. Any written plan should be available for inspection by the fire authority.

The purpose of the emergency plan is:

- to ensure that the people in your workplace know what to do to if there is a fire; and
- to ensure that the workplace can be safely evacuated.

In drawing up the emergency plan, you need to take the results of your risk assessment into account.

For most workplaces it should be easy to prepare a reasonable and workable emergency plan. In some small workplaces the final result may be some simple instructions covering the above points on a Fire Action Notice. However, in large or complex workplaces, the emergency plan will probably need to be more detailed.

If your workplace is in a building which is shared with other employers or occupiers, the emergency plan should be drawn up in consultation with those employers and the owner(s) or other people who have any control over any part of the building. It can help to agree on one person to co-ordinate this.

Part 2 - The five steps of fire risk assessment

Your plan should provide clear instructions on:

- the action employees should take if they discover a fire;
- how people will be warned if there is a fire;
- how the evacuation of the workplace should be carried out;
- where people should assemble after they have left the workplace and procedures for checking whether the workplace has been evacuated;
- identification of key escape routes, how people can gain access to them and escape from them to places of safety;
- the fire-fighting equipment provided;
- the duties and identity of employees who have specific responsibilities in the event of a fire;
- arrangements for the safe evacuation of people identified as being especially at risk, such as contractors, those with disabilities, members of the public and visitors;
- where appropriate, any machines/processes/power supplies which need stopping or isolating in the event of fire;
- specific arrangements, if necessary, for high-fire-risk areas of the workplace;
- how the fire brigade and any other necessary emergency services will be called and who will be responsible for doing this;
- procedures for liaising with the fire brigade on arrival and notifying them of any special risks, eg the location of highly flammable materials; and
- what training employees need and the arrangements for ensuring that this training is given.

If you have a larger or more complex workplace, then it might be helpful to you to include a simple line drawing. This can also help you check your fire precautions as part of your ongoing review. The drawing could show:

- essential structural features such as the layout of the workplace, escape routes, doorways, walls, partitions, corridors, stairways etc (including any fire-resisting structure and self-closing fire doors provided to protect the means of escape);
- means for fighting fire (details of the number, type and location of the fire-fighting equipment;
- the location of manually operated fire alarm call points and control equipment for the fire alarm;
- the location of any emergency lighting equipment and any exit route signs;
- the location of any automatic fire-fighting system and sprinkler control valve; and
- the location of the main electrical supply switch, the main water shut-off valve and, where appropriate, the main gas or oil shut-off valves.

Information and instructions for employees

It is important that your employees know how to prevent fires and what they should do if a fire occurs. They should all be given information about the fire precautions in the workplace and what to do in the event of a fire. You also need to ensure that you include employees working in the premises outside normal hours, such as cleaners or shift workers.

Part 2 - The five steps of fire risk assessment

Ensure that training and written information is given in a way that employees can understand, and take account of those with disabilities such as hearing or sight impairment, those with learning difficulties and those who do not use English as their first language.

On their first day, all employees should be given information about:

- the location and use of the escape routes from where they are working; and
- the location, operation and meaning of the fire warning system where they are working.

Fire Action Notices complement this information and should be prominently posted in key locations throughout the workplace. However, they are not a substitute for formal training.

Fire Action Notice

Note: The Fire Action Notice may also incorporate a simple plan indicating the route to a safe place. Where appropriate, the notice should include a translation into other languages.

Part 2 - The five steps of fire risk assessment

Training employees

The type of training should be based on the particular features of your workplace and:

- should explain your emergency procedures;
- take account of the work activity, the duties and responsibilities of employees;
- take account of the findings of the risk assessment; and
- be easily understandable by your employees.

You should ensure that all employees (and contractors) are told about the evacuation arrangements and are shown the means of escape as soon as possible after attending your premises.

Training should be repeated as necessary (usually once or twice a year) so that your employees remain familiar with the fire precautions in your workplace and are reminded about what to do in an emergency - including those who work in the premises outside normal hours, such as cleaners or shift-workers. It is very important that you tell your employees about any changes to the emergency procedures before they are implemented.

Training should preferably include practical exercises, eg fire drills, to check people's understanding of the emergency plan and make them familiar with its operation. In small workplaces, this might consist of making sure that employees are aware of details of the Fire Action Notice.

Your training should include the following:

- the action to take on discovering a fire;
- how to raise the alarm and what happens then;
- the action to take upon hearing the fire alarm;
- the procedures for alerting members of the public and visitors including, where appropriate, directing them to exits;
- the arrangements for calling the fire brigade;
- the evacuation procedures for everyone in your workplace to reach an assembly point at a safe place;
- the location and, when appropriate, the use of fire-fighting equipment;
- the location of the escape routes, especially those not in regular use;
- how to open all escape doors, including the use of any emergency fastenings;
- the importance of keeping fire doors closed to prevent the spread of fire, heat and smoke;
- where appropriate, how to stop machines and processes and isolate power supplies in the event of fire;
- the reason for not using lifts (except those specifically installed or adapted for evacuation of disabled people, see 'Use of lifts as means of escape' on page 70; and
- the importance of general fire safety and good housekeeping.

Part 2 - The five steps of fire risk assessment

In addition to the training in general fire precautions, employees should be informed of the risks from flammable materials used or stored on the premises. They should also be trained in the precautions in place to control the risks, particularly their role in reducing and controlling sources of ignition and fuel for the fire. Those working in high-risk areas should receive specific training in safe operating procedures and emergency responses. Where appropriate, training should cover:

- standards and work practices for safe operation of plant and equipment and safe handling of flammable materials (especially flammable liquids);
- housekeeping in process areas;
- reporting of faults and incidents, including leaks and spills of flammable liquids;
- emergency procedures for plant or processes in the event of fire, spills or leaks; and
- relevant legal requirements.

Further guidance on training is contained in the Approved Code of Practice to the Management of Health and Safety at Work Regulations 1992 (see the References section).

All the employees identified in your emergency plan who have a supervisory role in the event of fire (eg heads of department, fire marshals or wardens and, in some large workplaces, fire-fighting teams), should be given details of your fire risk assessment and receive additional training. This might include some or all of the measures listed at the beginning of this section.

STEP 5 - REVIEW AND REVISE

Sooner or later you may introduce changes in your workplace which have an effect on your fire risks and precautions, eg changes to the work processes, furniture, plant, machinery, substances, buildings, or the number of people likely to be present in the workplace. Any of these could lead to new hazards or increased risk. So if there is any significant change, you will need to review your assessment in the light of the new hazard or risk.

Do not amend your assessment for every trivial change or for each new job, but if a change or job introduces significant new hazards you will want to consider them and do whatever you need to keep the risks under control. In any case, you should keep your assessment under review to make sure that the precautions are still working effectively.

If a fire or 'near miss' occurs, then your existing assessment may be out of date or inadequate and you should reassess. It is a good idea to identify the cause of any incident and then review your fire risk assessment in the light of this.

PART 3 - FURTHER GUIDANCE ON FIRE PRECAUTIONS

REDUCING FIRE RISKS THROUGH GOOD MANAGEMENT

It helps to have a fire safety policy for your workplace which promotes good housekeeping and reduces the possibility of a fire occurring. Carelessness and neglect not only make the outbreak of a fire more likely but will inevitably create conditions which may allow a fire to spread more rapidly.

Step 3 in Part 2 of this guide listed various sources of ignition and flammable materials commonly found in workplaces. You were also introduced to measures and precautions which you could consider when evaluating the fire risk and considering improvements. This section gives further guidance on these measures which you may wish to consider implementing in order to reduce the risk of and from fire in your workplace.

Example of poor housekeeping

More guidance on particular fire hazards and precautions for specific industries can be found in Health and Safety Executive (HSE) guidance documents (see the References section).

Maintenance of plant and equipment

Plant and equipment which is not properly maintained can cause fires. The following circumstances often contribute to fires:

- poor housekeeping, such as allowing ventilation points on machinery to become clogged with dust or other materials - causing overheating;
- frictional heat (caused by loose drive belts, bearings which are not properly lubricated or other moving parts);
- electrical malfunction;
- flammable materials used in contact with hot surfaces;
- leaking valves or flanges which allow seepage of flammable liquids or gases; and
- static sparks (perhaps due to inadequate electrical earthing).

You may need to put a planned maintenance programme in place to make sure plant and other equipment is properly maintained (or review your programme if you already have one).

Storage and use of flammable materials

Workplaces in which large amounts of flammable materials are displayed, stored or

Part 3 - Further guidance on fire precautions

used can present a greater hazard than those where the amount kept is small. Wherever possible:

- quantities of flammable materials should be reduced to the smallest amount necessary for running the business and kept away from escape routes;
- highly flammable materials should be replaced by less flammable ones;
- remaining stocks of highly flammable materials should be properly stored outside, in a separate building, or separated from the main workplace by fire-resisting construction;
- employees who use flammable materials should be properly trained in their safe storage, handling and use; and
- stocks of office stationery and supplies and flammable cleaners' materials should be kept in separate cupboards or stores - if they open onto a corridor or stairway escape route, they should be fire-resisting with a lockable or self-closing fire door.

Flammable liquids

Flammable liquids can present a significant risk of fire. Vapours evolved are usually heavier than air and can travel long distances, so are more likely to reach a source of ignition. Liquid leaks and evolution of vapours can be caused by faulty storage (bulk and containers), plant and process - design, installation, maintenance or use.

Ignition of the vapours from flammable liquids remains a possibility until the concentration of the vapour in the air has reduced to a level which will not support combustion.

Detailed advice on the storage of flammable liquids is given in the HSE guidance documents listed in the References section. However, the following principles should be considered:

- The quantity of flammable liquids in workrooms should be kept to a minimum, normally no more than a half-day's or half a shift's supply.

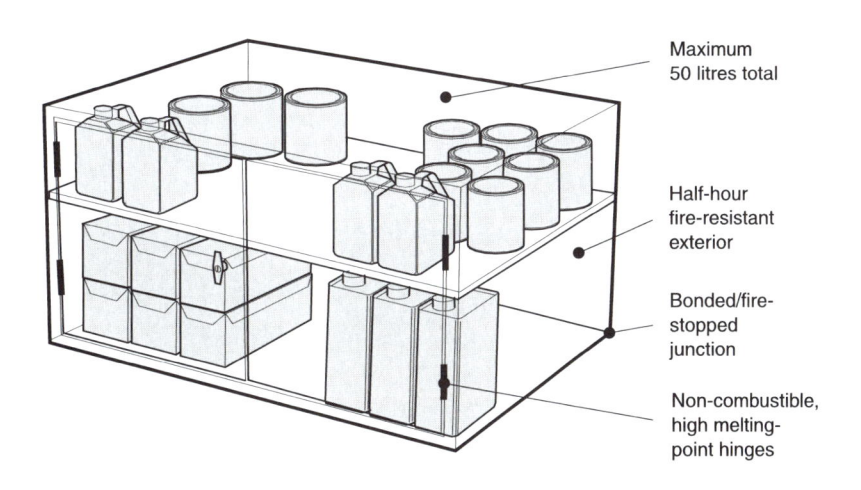

Maximum 50 litres total

Half-hour fire-resistant exterior

Bonded/fire-stopped junction

Non-combustible, high melting-point hinges

29

Part 3 - Further guidance on fire precautions

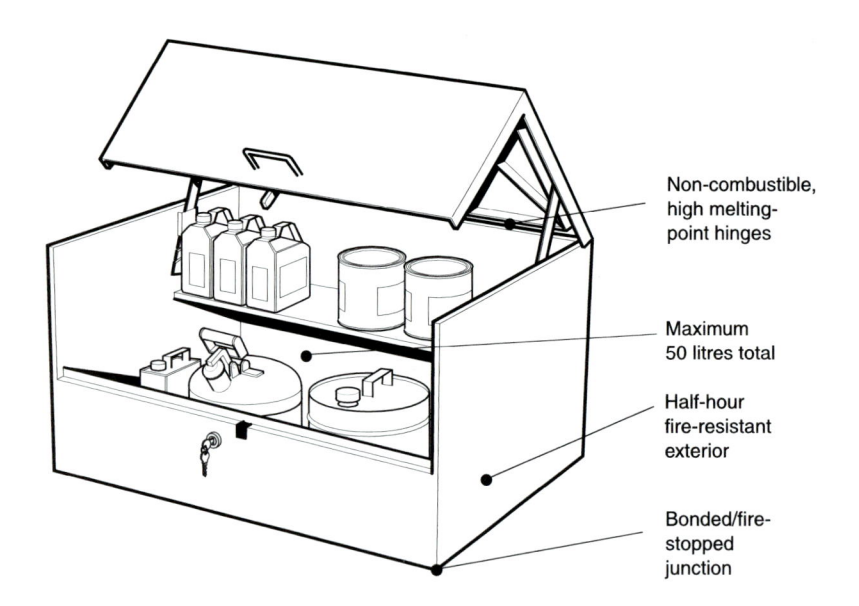

Storage in the workroom

- Flammable liquids, including empty or part-used containers, should be stored safely. Up to 50 litres of highly flammable liquids can be stored in the workroom if in closed containers in a fire-resisting (eg metal), bin or cabinet fitted with means to contain any leaks.
- Quantities greater than 50 litres should be stored in a properly designated store, either in the open air (on well ventilated, impervious ground, away from ignition sources) or in a suitably constructed storeroom.

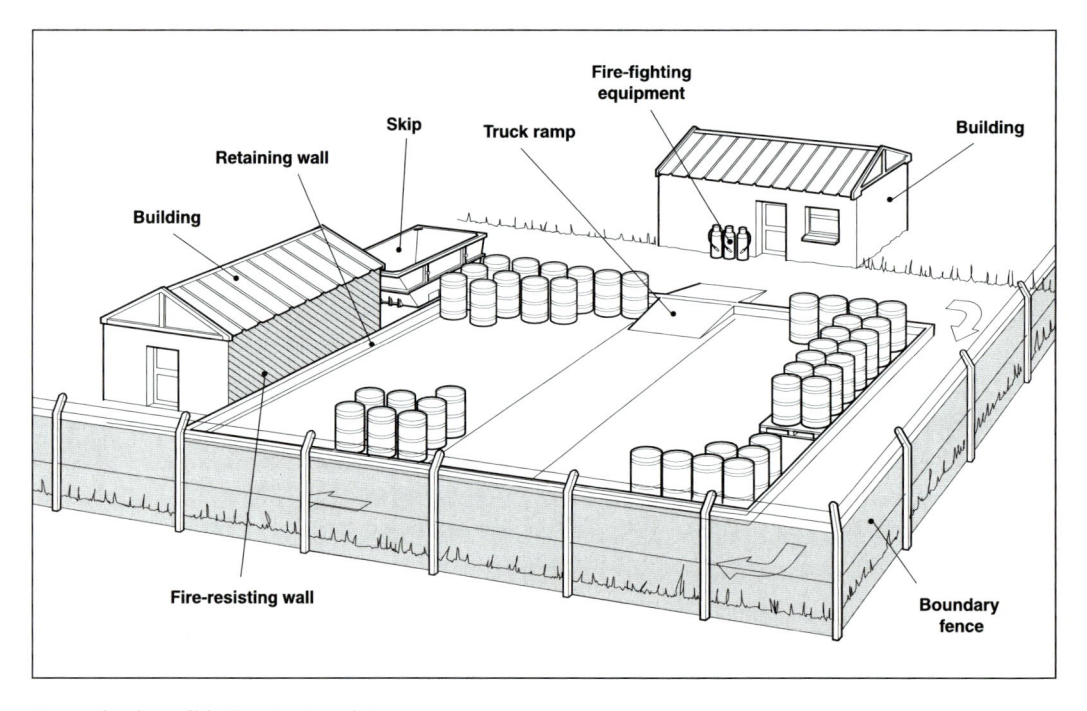

Example of a well laid out external storage area

Part 3 - Further guidance on fire precautions

- Where large quantities of flammable liquids are used they should, where possible, be conveyed by piping them through a closed system. Where a connection in such a system is frequently uncoupled and remade, a sealed-end coupling device should be used.
- Flammable liquids should not be dispensed within the store. Dispensing should take place in a well ventilated area set aside for this purpose, with appropriate facilities to contain and clear up any spillage.
- Container lids should always be replaced after use, and no container should ever be opened in such a way that it cannot be safely resealed.
- Flammable liquids should be stored and handled in well ventilated conditions. Where necessary, additional properly designed exhaust ventilation should be provided to reduce the level of vapour concentration in the air.
- Storage containers should be kept covered and proprietary safety containers with self-closing lids should be used for dispensing and applying small quantities of flammable liquids.
- Rags and cloths which have been used to mop up or apply flammable liquids should be disposed of in metal containers with well fitting lids and removed from the workplace at the end of each shift or working day.

Examples of special-purpose containers for flammable liquids

Example of metal container for cloths contaminated with flammable solvents

- There should be no potential ignition sources in areas where flammable liquids are used or stored and flammable concentrations of vapour may be present at any time. Any electrical equipment used in these areas, including fire alarm and emergency lighting systems, needs to be suitable for use in flammable atmospheres.

Part 3 - Further guidance on fire precautions

Work processes involving heat

You need to take special care if heat is used in conjunction with flammable materials, such as when cooking with fats. Ducts serving food grinders, cookers and ovens should be kept clean to avoid a build-up of grease.

Gas- and oil-burning plant, including fuel storage tanks where appropriate, should be installed in accordance with the appropriate standards. Such plant should be properly operated and maintained in accordance with the manufacturer's instructions. In particular, emergency fuel cut-off devices should be periodically checked to ensure they work and flues inspected regularly and cleaned as necessary.

Hot work

Activities such as welding, flame cutting, use of blow lamps or portable grinding equipment can pose a serious fire hazard and need to be strictly controlled when carried out in areas near flammable materials. This can be done by having a written permit to work for the people involved (whether they are your employees or those of a contractor).

A permit to work is appropriate in situations of high hazard/risk and, for example, where there is a need to:

- ensure that there is a formal check confirming that a safe system of work is being followed;
- co-ordinate with other people or activities;
- provide time-limits when it is safe to carry out the work; and
- provide specialised personal protective equipment (such as breathing apparatus) or methods of communication.

Any employees or contractors employed to carry out hot work should know that they cannot begin work until the person issuing the permit to work has explained the safety precautions fully. Hand-over of the permit should be recorded - usually by both the person issuing the permit and the person receiving it signing it.

Any location where 'hot work' is to take place should be examined to make sure that all material which could be easily ignited has either been removed or has been suitably protected against heat and sparks.

Suitable fire extinguishers should be readily available and a check made to ensure that people carrying out the work know how to use them and how to raise the alarm.

Where automatic fire detection equipment is installed and is likely to be actuated by heat, smoke or dust etc produced from the hot work, the detectors should be isolated for the duration of the work and reinstated immediately after work is finished.

Part 3 - Further guidance on fire precautions

1 Permit title

2 Permit number. Reference to other relevant permits or isolation certificates

3 Job location

4 Plant identification

5 Description of work to be done and its limitations

6 Hazard identification - including residual hazards and hazards introduced by the work

7 Precautions necessary - person(s) who carries out precautions, eg isolations, should sign that precautions have been taken

8 Protective equipment

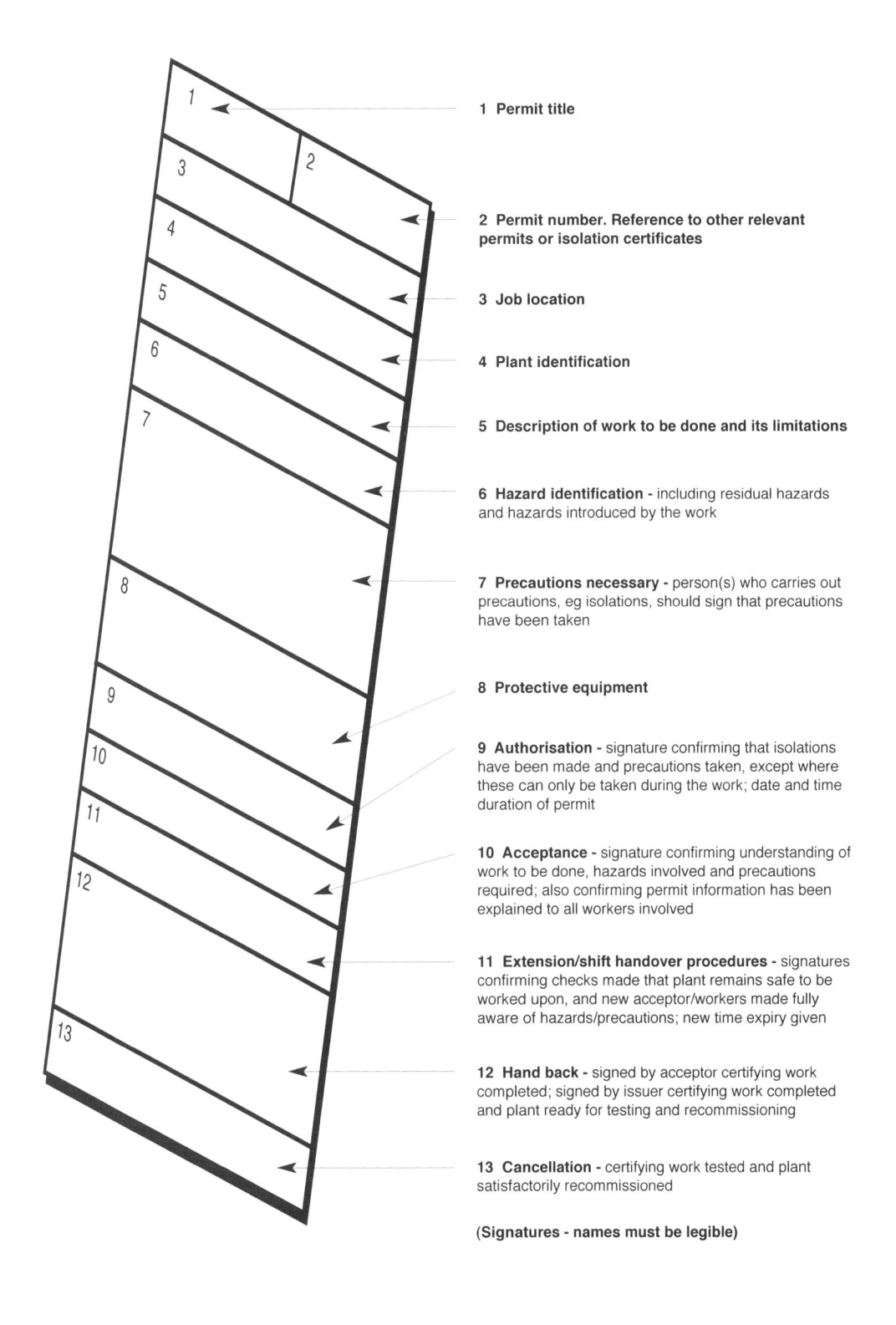

9 Authorisation - signature confirming that isolations have been made and precautions taken, except where these can only be taken during the work; date and time duration of permit

10 Acceptance - signature confirming understanding of work to be done, hazards involved and precautions required; also confirming permit information has been explained to all workers involved

11 Extension/shift handover procedures - signatures confirming checks made that plant remains safe to be worked upon, and new acceptor/workers made fully aware of hazards/precautions; new time expiry given

12 Hand back - signed by acceptor certifying work completed; signed by issuer certifying work completed and plant ready for testing and recommissioning

13 Cancellation - certifying work tested and plant satisfactorily recommissioned

(Signatures - names must be legible)

The essential requirements of a permit to work form

Part 3 - Further guidance on fire precautions

The permit to work should therefore contain the following details:

- measures to make sure all flammable material has been removed from the work area or, if it cannot be removed, adequately protected from heat or sparks;
- the fire-fighting equipment to be available in the work area;
- the permitted time span of the activity and the level of supervision required; and
- the actions to be taken when the work is finished, including initial and subsequent checks that there are no smouldering or hot materials which could allow a fire to break out at a later time.

Cylinders of flammable gases and oxygen should not be taken into confined spaces because of the risk of serious fire or explosion from a build-up of fuel gases, eg from a leak. All hot work equipment should be removed from the confined space whenever work stops - even for a break.

Electrical equipment

The main causes of fires originating from the use of electrical equipment are:

- overheating cables and electrical equipment due to overloading;
- damaged or inadequate electrical insulation on cables or wiring;
- flammable materials being placed too close to electrical equipment which may give off heat when operating normally or become hot due to a fault;
- arcing or sparking by electrical equipment; and
- the use of inappropriate or unsafe electrical equipment in areas where flammable atmospheres might be present, such as flammable liquid stores.

Explosions can occur if switchgear, power cables or motors are subject to a flow of electrical current which exceeds the maximum they were designed to work with.

All electrical systems must be designed, installed and maintained to prevent placing people in danger. There are a number of British Standards which offer guidance on how electrical systems and electrical equipment should be constructed and maintained. British Standard 7671 also offers practical advice on systems operating at up to 1000 V (see the References section for details).

Only suitably trained/qualified people should be allowed to install, maintain or otherwise work on electrical systems or equipment.

Further specific guidance on the use of electrical equipment and systems is given in HSE's *Memorandum of guidance on the Electricity at Work Regulations 1989* (see the References section).

Part 3 - Further guidance on fire precautions

Heating appliances

Make sure that individual heating appliances, particularly those which are portable, are used safely. Common causes of fire include:

- failing to follow the manufacturer's instructions when using or changing cylinders of liquefied petroleum gas (LPG);
- placing flammable materials on top of heating appliances;
- placing portable heaters too close to flammable materials; and
- careless refilling of heaters using paraffin.

Supplementary heating used during power failures or in exceptionally cold weather should be checked before being used and regularly serviced. If you use such appliances on a regular basis, it is better to use fixed convector heaters rather than portable heaters.

Smoking and the provision of ashtrays

It is better to allow people to smoke in places specifically set aside for that purpose rather than attempting to ban smoking in the workplace entirely. This can help to avoid unauthorised smoking in hidden or unsupervised areas such as store cupboards; this has led to serious fires. However, you should identify those areas where it is unsafe to smoke because there are materials which can be easily ignited. These areas should be clearly marked as no smoking areas.

The careless disposal of smokers' materials is one of the main causes of fire. Make sure that metal waste bins, ashtrays etc are provided in areas where smoking is permitted, and that these are emptied regularly. Ashtrays should not be emptied into containers which can be easily ignited; nor should their contents be disposed of with general rubbish.

Building and maintenance work

Many serious fires occur during building and maintenance work. This type of activity can increase the risk of fire and therefore needs to be carefully monitored and controlled (see also 'Hot work', on page 32). Extra fire precautions may be needed.

According to the size and use of the workplace and the nature of the work to be carried out, it may be necessary to carry out a new risk assessment so that all the hazards created by the work are identified and plans put in place to control the risks. Particular attention should be paid to:

- accumulations of flammable waste and building materials;
- the obstruction or loss of exits and exit routes;
- fire doors propped or wedged-open;
- openings created in fire-resisting partitions; and
- the introduction of extra electrical equipment or other sources of ignition.

Part 3 - Further guidance on fire precautions

At the beginning of the working day, it is essential to ensure that sufficient escape routes remain available for people in the workplace, whether employees or contractors, and that other fire safety arrangements are still effective. At the end of the working day, a check should be made to ensure that all risks of fire have been removed or adequately controlled.

Flammable materials used during construction or maintenance work, such as adhesives, cleaning materials or paints, should be securely stored in a well ventilated area when not in use and kept separate from other materials. Rooms in which they are used should be well ventilated and free from sources of ignition. Gas cylinders not in use should be stored securely outside the workplace, preferably in the open air. Smoking and the use of naked flames should not be allowed when using flammable materials.

In workplaces fitted with automatic fire detection systems, you need to consider how false alarms can be prevented during building or maintenance work, or where hot work is being undertaken, while maintaining adequate fire warning arrangements. At the end of such work the systems should be reinstated and tested (if they have been de-activated). Take special care when restoring gas and electricity supplies to ensure that equipment has not been inadvertently left on. Further information on fire precautions on construction sites can be found in the publications referred to in the References section.

Flammable rubbish and waste

Flammable rubbish and waste should not be stored, even as a temporary measure, in escape routes such as corridors, stairways or lobbies, or where it can come into contact with potential sources of heat. Accumulations of flammable rubbish and waste in the workplace should be avoided, removed at least daily and suitably stored away from the building.

Part 3 - Further guidance on fire precautions

Do not allow flammable waste, unused materials, and redundant packaging, such as cardboard, wooden or plastic containers and wooden pallets, to build up at the workplace; these must be safely stored until they are removed from your premises. Where a skip is provided for the collection of debris or rubbish, it should be positioned so that a fire in it will not put the workplace, or any other structure, at risk.

Parts of the workplace which are not normally occupied, such as basements, store rooms and any area where a fire could grow unnoticed, should be regularly inspected and cleared of non-essential flammable materials and substances. You should also protect such areas against entry by unauthorised people.

If the workplace has waste or derelict land on or bordering it, you should keep any undergrowth under control (using a non-flammable weedkiller if necessary) so that a fire cannot spread through dry grass, for example. If you do have to burn bonfires in yards or other open areas, they should be carefully controlled and in positions where they will not pose a threat to the workplace. You should make sure that any bonfire is completely out before closing the workplace for the day.

Reducing the risk of arson

Deliberately started fires pose very significant risks to all types of workplace. A study conducted by the Home Office (*Safer communities: towards effective arson control*) has suggested that the cost of arson to society as a whole has now reached over £1.3 billion a year. The same study suggests that, in an average **week**, arson results in:

- 3500 deliberately started fires;
- 50 injuries;
- two deaths; and
- a cost to society of at least £25 million.

The possibility of arson should be considered as a component of your risk assessment and it is one that you can do much to control. The majority of deliberately started fires occur in areas with a known history of vandalism or fire-setting. Typically, local youths light the fires outside the premises as an act of vandalism, using flammable materials found nearby. Appropriate security measures, including the protection of stored materials and the efficient and prompt removal of rubbish, can therefore do much to alleviate this particular problem.

There is a joint duty on local authorities and police to co-operate with other organisations (including fire authorities) to formulate and implement a strategy to reduce crime and disorder (including arson) in their local area. You should therefore seek advice from the local police or the fire authority who will involve the other agencies as appropriate. The Arson Prevention Bureau (see Further information) can provide further guidance on arson prevention measures for a range of building types.

Part 3 - Further guidance on fire precautions

Occasionally, arson attacks in the workplace are committed by employees or ex-employees. Employers and other workers should be aware of this potential threat and be alert for early signs, such as a series of unexplained small fires. Again, the police, fire authority or the Arson Prevention Bureau can provide further useful guidance.

FIRE DETECTION AND WARNING

A fire in your workplace must be detected quickly and a warning given so that people can escape safely. Early discovery and warning will increase the time available for escape and enable people to evacuate safely before the fire takes hold and blocks escape routes or makes escape difficult.

The nature and extent of the fire detection and warning arrangements in your workplace will need to satisfy the requirements indicated by your risk assessment.

Fire detection

All workplaces should have arrangements for detecting fire. During working hours, fires are often detected through observation or smell, and for many workplaces automatic fire detection equipment may not be needed.

However, you need to think about any parts of the workplace where a fire could start and spread undetected. This could be a storage area or a basement that is not visited on a regular basis, or a part of the workplace that has been temporarily vacated, for example at mealtimes. Fires that start and develop unnoticed can pose a serious danger to people in the workplace.

The usual method of protecting people in workplaces where fire could develop for some time before being discovered is to protect vital escape routes, particularly staircase routes, with fire-resisting construction which may include fire-resisting doors.

Installing an effective, reliable automatic fire detection system, linked to an effective fire warning system, can sometimes allow people to reassess the degree of structural fire protection required on escape routes. This can provide a more cost-effective and convenient fire precaution. However, the whole subject of trade-offs between structural protection and other fire protection systems is a complex one and such decisions should only be made after consultation with your local fire authority.

In some workplaces, such as those providing sleeping accommodation or care facilities, automatic fire detection and a high degree of structural protection are essential in ensuring a satisfactory standard of fire safety.

In small workplaces, it may be unnecessary to provide a sophisticated automatic fire detection system based on point-type fire detectors linked via control equipment to separate fire warning devices. In these cases, good quality, interlinked domestic smoke alarms (mains powered with battery back-up) could provide an automatic means of

Part 3 - Further guidance on fire precautions

detecting fire. Each of these units contains a fire detector and a warning device and can operate independently or in conjunction with any other unit to which it is interlinked.

In other situations, for example where the only escape route from a room is through an outer room where a fire may start unnoticed, a single smoke alarm of the same type as described in the previous paragraph, positioned in the outer room, can provide an early warning and allow workers to escape before their route is cut off. Smoke alarms should conform with the requirements of British Standard 5446: Part 1.

Such basic smoke alarms tend to be more sensitive than smoke detectors used in more sophisticated fire detection/alarm systems. You need to be aware of any potential problems unwanted fire signals may cause. In some cases, unwanted fire signals can be reduced by using optical smoke alarms rather than ionisation ones.

This simple but effective way of providing automatic fire detection could provide a cost-effective solution to difficult situations where early warning is vital in ensuring the safe evacuation of employees. However, smoke alarms designed for domestic use are usually manufactured to different standards from those for automatic fire detection systems. The resulting reliability may therefore be lower and such smoke alarms may not be appropriate for your workplace, depending on the processes involved.

Whichever type of system you use, the detector type chosen should be appropriate for the premises to be protected, for example, a heat detector may function better than a smoke detector in a fume-laden or dusty environment but may not be appropriate for the rest of the protected premises. Choosing the right type of detector will reduce the chances of it giving false fire signals. False alarms can cause costly interruptions to manufacturing processes and business activities. They also increase the risk to occupants if the fire brigade is responding to a false fire call and is not so readily available to tackle a real fire.

Before installing an automatic fire detection system or a series of interlinked smoke alarms, it is advisable to consult the fire authority about what you propose. This can help make sure the system is appropriate to the circumstances of the workplace and avoid unnecessary costs.

Automatic fire detectors or smoke alarms do not remove the need to provide a means for people to manually raise a fire warning, and this will be essential in the majority of workplaces.

Fire warning

In workplaces that are only small buildings or small open areas, the means of raising the alarm may be simple. For instance, where all employees work near to each other, a shouted warning 'Fire!' by the person discovering the fire may be all that is needed. But you will need to be satisfied that the warning can be heard and understood throughout the workplace, including the toilets.

Part 3 - Further guidance on fire precautions

Where employees are dispersed more widely and it cannot be guaranteed that a shouted warning will be heard, a manually operated sounder (such as a rotary gong or handbell) or a simple manual call point, combined with a bell, battery and charger, may be suitable. However, you must ensure that any manually operated system is positioned so that it can be reached by the person discovering a fire and then operated for sufficient time to alert everyone in the workplace, without exposing the operator to danger.

In larger buildings, a suitable electrically operated fire warning system, with manual call points positioned both on exit routes and adjacent to final exits, should be installed. This should have sufficient sounders for the warning to be clearly heard throughout the workplace. The sound used as a fire warning should be distinct from other sounds in the workplace and, where background noise levels are high or an employee has a hearing impairment, it may also be necessary to install a visual alarm such as a distinctive flashing or rotating light.

In more complex buildings such as retail premises, where the evacuation system is based on staged or phased evacuation (see the 'Means of escape in case of fire' section), or where people are unfamiliar with the fire warning arrangements, you might consider installing a voice evacuation system. The system could form part of a public address system and could give both fire warning signals and verbal instructions in the event of fire.

Where a public address system is used in conjunction with a fire warning system, both should over-ride any other function of the equipment (such as playing music). The public address element of the system should give clear verbal instructions and should over-ride the fire warning signal - this should be distinct from other signals which may be in general use.

In workplaces covering large areas, using a public address-based warning system for people inside, and a radio-telephone system or walkie-talkie for people outside, can be an effective way of supplementing a conventional fire warning system. This would allow clarification of the precise nature and location of the emergency, and instructions on the pre-determined action to be given.

If an automatic fire detection system and a manually operated electrical alarm system are installed in the same workplace, they should normally be incorporated into a single integral system. Voice evacuation systems should be similarly integrated to prevent confusion.

Electrical fire detection and alarm systems should normally comply with British Standard 5839: Part 1, voice alarm systems should comply with British Standard 5839: Part 8. Although intended to cover domestic dwellings, British Standard 5839: Part 6 can offer useful information about systems which may be considered appropriate for use in some workplaces. Again, it is advisable to consult the fire authority about your proposals before installing a new fire warning system or altering an existing one.

Part 3 - Further guidance on fire precautions

The Health and Safety (Signs and Signals) Regulations 1996 require that, in a workplace, fire safety signs and signals requiring some form of power (mains-powered smoke alarms and other fire warning systems) must be provided with a guaranteed emergency supply in the event of a power cut.

MEANS OF ESCAPE IN CASE OF FIRE

The principle on which means of escape provisions are based is that the **time available for escape** (an assessment of the length of time between the fire starting and it making the means of escape from the workplace unsafe) is greater than the **time needed for escape** (the length of time it will take everyone to evacuate once a fire has been discovered and warning given).

Regardless of the location of a fire, once people are aware of it they should be able to proceed safely along a recognisable escape route, to a place of safety. In order to achieve this, it may be necessary to protect the route, ie by providing fire-resisting construction. A protected route may also be necessary in workplaces providing sleeping accommodation or care facilities.

The means of escape is likely to be satisfactory if your workplace is fairly modern and has had building regulation approval or if it has been found satisfactory following a recent inspection by the fire authority (and in each case you have not carried out any significant material or structural alterations or made any change to the use of the workplace). However, you should still carry out a risk assessment to ensure that the means of escape remains adequate.

If, as a result of your risk assessment, you propose making any changes to the means of escape, you should consult the fire authority (in Scotland you must seek the agreement of the building control authority) before making any changes.

When assessing the adequacy of the means of escape you will need to take into account:

- the findings of your fire risk assessment;
- the size of the workplace, its construction, layout, contents and the number and width of the available escape routes;
- the workplace activity, where people may be situated in the workplace and what they may be doing when a fire occurs;
- the number of people who may be present, and their familiarity with the workplace; and
- their ability to escape without assistance.

Part 3 - Further guidance on fire precautions

> **Technical terms relating to means of escape**

There are a number of technical terms used in this section which are defined as follows:

Compartment: A part of a building separated from all other parts of the same building by fire-resisting walls, ceilings and floors.

Emergency escape lighting: That part of the emergency lighting system provided for use when the electricity supply to the normal lighting fails so as to ensure that the means of escape can be safely and effectively used at all times.

Final exit: The end of an escape route from a workplace giving direct access to a place of safety such as a street, walkway or open space, and located to ensure that people can disperse safely from the vicinity of the workplace and the effects of fire.

Fire door: A door assembly which, if tested under the relevant British Standard (see the References section), would satisfy the criteria for integrity for at least 20 minutes or a longer period if this is specified.

Fire-resisting (fire-resistance): The ability of a component or construction of a building to satisfy, for a stated period of time, some or all of the appropriate criteria specified in the relevant British Standard (see the References section).

Place of safety: A place beyond the building in which a person is no longer in danger from fire.

Protected route: A route with an adequate degree of fire protection including walls (except external walls), doors, partitions, ceilings and floors separating the route from the remainder of the building.

Protected lobby: A fire-resisting enclosure separated from other parts of the workplace by self-closing fire doors, leading by a second set of self-closing fire doors to a protected stairway with no other openings, other than from toilets (which contain no fire risk) or lifts.

Protected stairway: A stairway which is adequately protected from fire in the adjoining accommodation by fire-resisting construction and either leads to a final exit or along a protected route to a final exit.

Storey exit: An exit people can use so that, once through it, they are no longer at immediate risk. This includes a final exit, an exit to a protected lobby or stairway (including an exit to an external stairway) and an exit provided for means of escape through a compartment wall through which a final exit can be reached.

Part 3 - Further guidance on fire precautions

Arrangements for evacuating the workplace

You need to consider how you will arrange the evacuation of the workplace in the light of your risk assessment and the other fire precautions you have or intend to put in place. These arrangements will form an integral part of your emergency plan and must be included in the instruction and training you need to give your employees.

You must take account not only of the people in your workplace (employed or otherwise) who may be able to make their own escape, but also those who may need assistance to escape, eg by having adequate staffing levels in premises providing treatment or care.

In most workplaces, the evacuation in case of fire will simply be by means of everyone reacting to the warning signal given when the fire is discovered and making their way, by the means of escape, to a place of safety away from the workplace. This is known as a 'simultaneous' evacuation and will normally be initiated by the sounding of the general alarm over the fire warning system.

In some larger workplaces, the emergency arrangements are designed to allow people who are not at immediate risk from a fire to delay starting their evacuation. It may be appropriate to 'stage' the evacuation by initially evacuating only the area closest to the fire and warning other people to 'stand by'. The rest of the people are then evacuated if it is necessary to do so. This is known as a 'two stage' evacuation. The fire warning system should be capable of giving two distinctly different signals (warning and evacuation) or give appropriate voice messages.

Alternatively, and usually only in the most complex buildings, the evacuation could take place floor by floor. This is normally done by immediately evacuating the floor where the fire is located and the floor above. The other floors are then evacuated one by one to avoid congestion on the escape routes. This is known as 'phased evacuation'. Because of the extra time this type of evacuation takes, other fire precautions are likely to be required. These include:

- voice alarm systems;
- fire control points;
- compartmentation of the workplace (with fire-resisting construction); and
- sprinklers in buildings where the top floor is 30 metres or more above ground level.

43

Part 3 - Further guidance on fire precautions

In some cases it may not be appropriate for a general alarm to start immediate evacuation. This could be because of the number of members of the public present and the need for employees to put pre-arranged plans for the safe evacuation of the workplace into action. In such circumstances a 'staff alarm' can be given (by personal pagers, discreet sounders or a coded phrase on a public address system etc).

Following the staff alarm, a more general alarm signal can be given and a simultaneous, two stage or phased evacuation started (see 'Arrangements for evacuating the workplace'). The general alarm may be activated automatically if manual initiation has not taken place within a pre-determined time.

You should only plan to use staged or phased evacuation schemes, or a staff alarm system, if you have sought the advice of the fire authority and they have given their approval to the proposal.

Assessing means of escape

The aim of the following paragraphs is to provide enough information for you to make a reasonable assessment of the escape routes from your workplace to decide whether they are adequate and can be safely used in the event of fire.

Because of the wide variation in the type of workplaces covered by the Fire Regulations, it is only possible to give a general guide to the level of precautions required to satisfy those Regulations in most workplaces. So this guide does not seek to give specific advice about each individual type of workplace. If your workplace is unusual, particularly if it is a large, complex premises or involved with specialised activities or risks, you may wish to seek specialist advice or refer to further specific guidance (see the References section).

In some cases, it may be necessary to provide additional means of escape or to improve the fire protection of existing escape routes. If, having carried out your risk assessment, you think this might be the case in your workplace, consult the fire authority and, where necessary, your local building control officer before carrying out any alterations.

It would be a time-consuming and complicated process, requiring specialist expertise, to establish the time needed in each individual case. So this guide uses an established method for assessing means of escape which has been found to be generally acceptable in all but the most particular circumstances. This method is based upon limiting travel distances according to the category of potential fire risk the workplace falls into.

These distances ensure that people will be able to escape within the appropriate period of time. You can use actual calculated escape times but should do so only after consulting a fire safety specialist with appropriate training and expertise in this field.

Fire risk categories for assessing the means of escape

In general, most workplaces can be categorised as high, normal or low risk. Examples of the type of workplace or areas within workplaces likely to fall within these categories are:

Part 3 - Further guidance on fire precautions

High Where highly flammable or explosive materials are stored or used (other than in small quantities).

Where unsatisfactory structural features are present such as:

- lack of fire-resisting separation;
- vertical or horizontal openings through which fire, heat and smoke can spread;
- long and complex escape routes created by extensive subdivision of large floor areas by partitions, or the distribution of display units in shops or machinery in factories; and
- large areas of flammable or smoke-producing surfaces on either walls or ceilings.

Where permanent or temporary work activities are carried out which have the potential for fires to start and spread such as:

- workshops in which highly flammable materials are used, eg paint spraying;
- areas where the processes involve the use of naked flame, or produce excessive heat;
- large kitchens in works canteens and restaurants;
- refuse chambers and waste disposal areas; and
- areas where foamed plastics or upholstered furniture are stored.

Where there is a significant risk to life in case of fire, such as where:

- sleeping accommodation is provided for staff, the public or other visitors in significant numbers;
- treatment or care is provided where the occupants have to rely upon the actions of limited numbers of staff for their safe evacuation;
- there is a high proportion of elderly or infirm people, or people with temporary or permanent physical or mental disabilities, who need assistance to escape;
- groups of people are working in isolated parts of the premises such as basements, roof spaces, cable ducts and service tunnels etc; and
- large numbers of people are present relative to the size of the premises (eg sales at department stores) or in other circumstances where only a low level of assistance may be available in an emergency (eg places of entertainment and sports events).

Normal Where any outbreak of fire is likely to remain confined or only spread slowly, allowing people to escape to a place of safety.

45

Part 3 - Further guidance on fire precautions

> Where the number of people present is small and the layout of the workplace means they are likely to be able to escape to a place of safety without assistance.
>
> Where the workplace has an effective automatic warning system, or an effective automatic fire-extinguishing, -suppression or -containment system, which may reduce the risk classification from high risk.

Low   Where there is minimal risk to people's lives and where the risk of fire occurring is low, or the potential for fire, heat and smoke spreading is negligible.

The work you have done on assessing the risks and reducing the risk of fire occuring, together with the knowledge you have gained about the location of people at risk, should generally provide you with the information you need to establish the risk category or categories of your workplace.

General principles for escape routes

Other than in small workplaces, or from some rooms of low or normal fire risk, there should normally be alternative means of escape from all parts of the workplace.

Routes which provide means of escape in one direction only (from a dead-end) should be avoided wherever possible as this could mean that people have to move towards a fire in order to escape.

Each escape route should be independent of any other and arranged so that people can move away from a fire in order to escape.

Escape routes should always lead to a place of safety. They should also be wide enough for the number of occupants and should not normally reduce in width.

Escape routes and exits should be available for use and kept clear of obstruction at all times.

Evacuation times and length of escape routes

The aim is, from the time the fire alarm is raised, for everyone to be able to reach a place of relative safety, ie a storey exit (see 'Technical terms relating to means of escape'), within the time available for escape.

The time for people to reach a place of relative safety should include the time it takes them to react to a fire warning. This will depend on a number of factors including:

- what they are likely to be doing when the alarm is raised, eg sleeping, having a meal etc;

Part 3 - Further guidance on fire precautions

- what they may have to do before starting to escape, eg turn off machinery, help other people etc; and
- their knowledge of the building and the training they have received about the routine to be followed in the event of fire.

Where necessary, you can check these by carrying out a practice drill.

To ensure that the time available for escape is reasonable, the length of the escape route from any occupied part of the workplace to the storey exit should not exceed:

Where more than one route is provided

25 metres	-	high-fire-risk area;
32 metres	-	normal-fire-risk (sleeping) area;
45 metres	-	normal-fire-risk area;
60 metres	-	low-fire-risk area.

Where only a single escape route is provided

12 metres	-	high-fire-risk area;
16 metres	-	normal-fire-risk (sleeping) area;
18 metres	-	normal-fire-risk area (except production areas in factories);
25 metres	-	normal-fire-risk area (including production areas within factories);
45 metres	-	low-fire-risk area.

Where the route leading to a storey exit starts in a corridor with a dead-end, then continues via a route which has an alternative, the total distance should not exceed that given above for 'Where more than one route is provided'. However, the distances within the 'dead-end portion' should not exceed those given for 'Where only a single escape route is provided'. (See the diagram on page 49.)

People with disabilities

You may need to make special arrangements for staff with disabilities, which should be developed in consultation with the staff themselves. British Standard 5588: Part 8 gives guidance and provides full information (see the References section). There is further detailed guidance under 'Disabled people' later in Part 3.

Premises providing residential care and/or treatment

The distances shown in the paragraphs above may not be suitable for workplaces providing residential care. You should refer to the relevant guidance listed in the References section or seek specialist advice.

Part 3 - Further guidance on fire precautions

Number and width of exits

There should be enough available exits, of adequate width, from every room, storey or building. The adequacy of the escape routes and doors can be assessed on the basis that:

- a doorway of no less than 750 millimetres in width is suitable for up to 40 people per minute (where doors are likely to be used by wheelchair users the doorway should be at least 800 millimetres wide); and
- a doorway of no less than 1 metre in width is suitable for up to 80 people per minute.

Where more than 80 people per minute are expected to use a door, the minimum doorway width should be increased by 75 millimetres for each additional group of 15 people.

For the purposes of calculating whether the existing exit doorways are suitable for the numbers using them, you should assume that the largest exit door from any part of the workplace may be unavailable for use. This means that the remaining doorways should be capable of providing a satisfactory means of escape for everyone present.

Inner rooms

You should avoid situations where the only escape route for people in an inner room is through one other room (the access room). The exception to this is where the people in the inner room can be quickly made aware of a fire in the outer one and this is not an area of high fire risk. Inner rooms should generally not be used as sleeping accommodation. The diagram below shows the alternatives you can use to make sure people in an inner room know there is a fire.

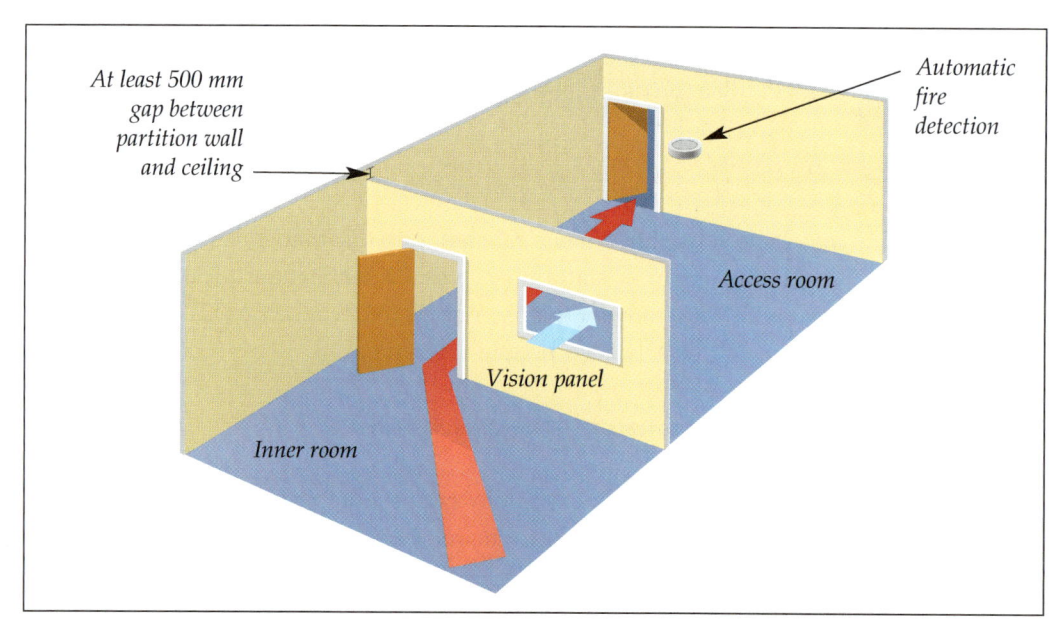

Example of alternative arrangements for ensuring that occupants of an inner room are aware of an outbreak of fire in the access room

Part 3 - Further guidance on fire precautions

Where there is no automatic fire detection system, it may be reasonable to provide a self-contained smoke alarm which is solely within the access room, as long as it is clearly audible within the inner room.

Corridors

Corridors should generally be about 1 metre wide, although wheelchair users will need a width of 1.2 metres. The doors should be aligned with the walls of the rooms so that the floor area is effectively divided into two or more parts. To avoid having to travel long distances in corridors affected by smoke, those corridors which are more than 30 metres long (45 metres in offices and factories) should be subdivided into approximately equal parts by providing, close-fitting, self-closing fire doors.

Where a corridor only leads in one direction, or serves sleeping accommodation, it should be constructed of fire-resisting partitions and self-closing fire doors (see the diagram below). This does not apply to toilets.

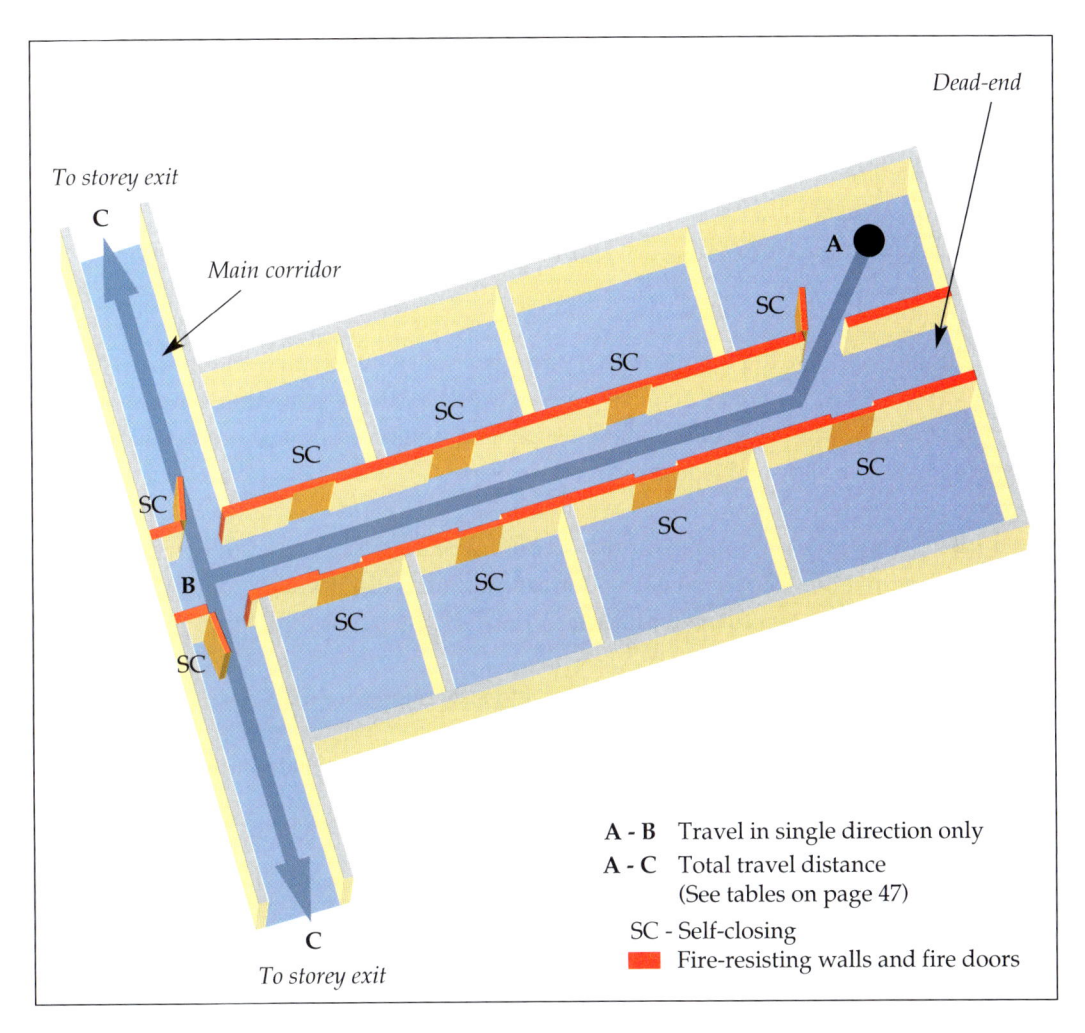

A - B Travel in single direction only
A - C Total travel distance
       (See tables on page 47)
SC - Self-closing
Fire-resisting walls and fire doors

Example of fire-resisting separation where a corridor from an initial dead-end meets a main corridor providing alternative means of escape

Part 3 - Further guidance on fire precautions

Stairways

A stairway should be of sufficient width for the number of people who are likely to use it in an emergency and it should not normally be less than 1 metre wide. However, a narrower one may be adequate if you are sure that only a few people, who are familiar with the stairway, will use it.

Where more than one stairway is provided, you should assume that the widest one may be unusable as a result of the fire. This means that the remaining stairway(s) will need to provide a satisfactory escape route for everyone present.

There may be no need for you to discount the widest stairway where each stairway is reached through a protected lobby. Certain other compensatory features, such as sprinklers or smoke control systems, may also be considered.

Stairways should normally be protected by fire-resisting partitions and fire-resisting, self-closing doors (except toilet doors) and lead directly to a way out of the building. An unprotected stairway may, however, be suitable in workplaces of low or normal fire risk, provided that:

- the stairway links no more than two floors and those floors are not linked to another floor by an unprotected stairway; and
- it is additional to that required for escape purposes; and
- no escape route from a dead-end situation on an upper floor passes the access to such a stairway.

People should not have to pass through a protected stairway to reach an alternative stairway. Where this cannot be achieved, a stairway may be by-passed, for instance by using doors connecting adjacent rooms. In such situations the doors should be kept free from obstruction and available for use at all times.

A single stairway may be suitable for means of escape in workplaces of low or normal fire risk, provided that people on each floor can reach it within the appropriate travel time (see details of distances earlier in this section). It also needs to:

- be constructed as a protected stairway and serve no more than three floors above, or one floor below, ground level;
- be accessed, other than at the top floor, by means of a protected lobby or protected corridor;
- be of sufficient width to accommodate the number of people who may need to use it in an emergency; and
- lead direct to open air.

There is no need for you to provide protected lobbies where the workplace is of low fire risk. This also applies to workplaces of normal fire risk, provided that either an automatic fire detection system or sprinklers linked directly into the fire alarm system are installed in the rooms or areas leading directly onto the protected stairway.

Part 3 - Further guidance on fire precautions

In small workplaces of low or normal fire risk, unprotected stairways (including a single stairway) may be satisfactory as a means of escape, provided that:

- the stairway provides access between the ground and first floor and/or ground floor and basement only, and an exit can be reached from any part of those floors within the escape times given for single escape routes earlier in this section; and
- access to the stairway is clearly visible from any part of the floor it serves and it exits not more than 6 metres from a storey exit leading direct to open air at ground level.

Where an external stairway is provided, any door or window (other than toilet windows) opening onto the stairway, or within 1.8 metres horizontally or 9 metres vertically of it, should be fire-resisting. Windows should be unopenable and doors should be self-closing.

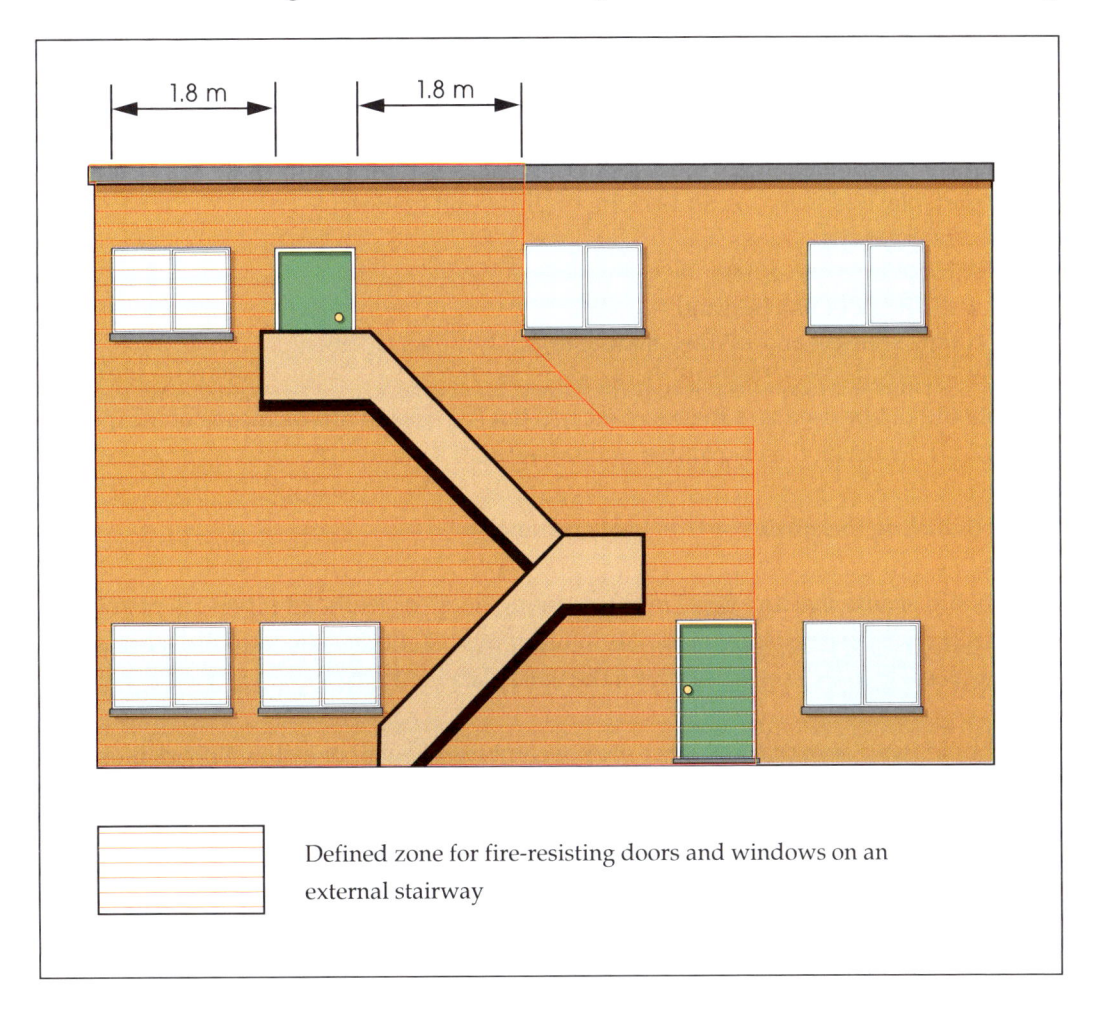

Defined zone for fire-resisting doors and windows on an external stairway

In exceptional circumstances, a small number of unprotected, openable windows may be allowed, provided that the rooms containing them are separated from the rest of the building by fire-resisting construction and the external stairway is not the only one from the upper storeys.

Part 3 - Further guidance on fire precautions

Accommodation stairways

Your workplace may have stairways that are not needed as part of the formal means of escape. These stairways are known as accommodation stairs and will not need to be protected, provided that:

- they do not serve protected corridors;
- they do not link more than two floors; and
- people do not have to move towards the head of an unprotected stairway to make their escape.

Means of escape for use by staff

The features listed below are not normally acceptable as a means of escape for members of the public as they are not conventional escape routes. However, they may, in certain circumstances, be used by a small number of staff if they are trained to do so or use the exit during their normal work activity:

- revolving doors (except those specifically designed for escape purposes);
- portable, foldaway, vertical-raking or throw-out ladders;
- window exits;
- wicket doors and gates;
- wall and floor hatches; and
- rolling shutters and folding, sliding or up-and-over doors.

Lifts should not be used as a means of escape, but see 'Use of lifts as means of escape' on page 70 for details on the evacuation of disabled staff.

Reducing the spread of fire, heat and smoke

You should ensure that any holes in fire-resisting floors or walls, eg pipework openings, are filled in with fire-resisting materials in order to prevent the spread of fire, heat and smoke. (See 'How fire spreads through the workplace' in Part 2.)

You should make sure that any large area of combustible wall or ceiling linings is either removed, treated or suitably covered to reduce the possibility of the linings contributing to the rapid spread of fire. Such linings must not be used in escape routes. (Information on the suitability of wall and ceiling linings can be found in the Home Office publication *Guide to the fire precautions in existing places of work that require a fire certificate. Factories, offices, shops and railway premises* - see the References section.)

Exhibitions and displays

Any exhibition or display with large amounts of flammable materials, such as paper, textiles and cotton wool, can allow fire to spread rapidly. You should therefore avoid using such materials wherever possible. Any permanent or semi-permanent displays, including wall displays, should ideally be placed behind glass.

Part 3 - Further guidance on fire precautions

Noticeboards should be kept as small as possible and should be fixed securely in position. They should not be continuous along the length of a wall, sited above heaters etc or contain an excessive amount of paper (eg overlapping or multi-sheet notices).

Defining the escape route

The contents of any room in which people are working or any open floor area to which the public are admitted should be arranged to ensure that there is a clear passageway to all escape routes. This may mean that you will need to clearly define the routes, for example by marking the floor or by providing a contrasting floor covering.

Items prohibited on an escape route

You should make sure that items which pose a potential fire hazard or those which could cause an obstruction are not located in corridors or stairways intended for use as a means of escape. In particular, the following items should not be located in protected routes, or in a corridor and stairwell which serves as the sole means of escape from the workplace, or part of it:

- portable heaters of any type;
- heaters which have unprotected naked flames or radiant bars;
- fixed heaters using a gas supply cylinder, where the cylinder is within the escape route;
- oil-fuelled heaters or boilers;
- cooking appliances;
- upholstered furniture;
- coat racks;
- temporarily stored items including items in transit, eg furniture, beds, laundry, waste bins etc;
- lighting using naked flames;
- gas boilers, pipes, meters or other fittings (except those permitted in the standards supporting the building regulations and installed in accordance with the 'Gas Safety Regulations');
- gaming and/or vending machines; and
- electrical equipment (other than normal lighting, emergency escape lighting, fire alarm systems, or equipment associated with a security system), eg photocopiers.

Escape doors

Doors people have to pass through in order to escape from the workplace should open in the direction of travel where:

- more than 50 people may have to use the door;
- the door is at or near the foot of a stairway;
- the door serves a high-fire-risk area (see 'Fire risk categories for assessing the means of escape' earlier in this section); or
- the door is on an exit route from a building used for public assembly, such as a place of public entertainment, a conference centre or exhibition hall.

53

Part 3 - Further guidance on fire precautions

You should make sure that people escaping can open any door on an escape route easily and immediately, without the use of a key. All outward opening doors used for means of escape, which have to be kept fastened while people are in the building, should be fitted with a single form of release device such as a panic latch, a panic bolt, or a push pad.

Where a door needs to be fastened by a security device, it should be the only fastening on the door and you will have to make sure that all your staff know how it works. Such devices are not normally suitable for use by members of the public. You should display a notice explaining the method of operation and, if necessary, provide a suitable tool so that the device can be operated safely.

Fire doors

Where fire doors are provided they should be fitted with effective self-closing devices and labelled 'Fire Door - Keep Shut'. Fire doors to cupboards and service ducts need not be self-closing, provided they are kept locked and labelled 'Fire Door - Keep Locked Shut'. (Signs should meet the requirements of British Standard 5499 - see the References section.)

Self-closing fire doors may be held open by automatic door release mechanisms which are either:

- connected into a manually operated electrical fire alarm system incorporating automatic smoke detectors in the vicinity of the door; or
- actuated by independent smoke detectors (not domestic smoke alarms) on each side of the door.

Where such mechanisms are provided, it should be possible to release them manually. The doors should be automatically closed by:

- the actuation of a smoke-sensitive device on either side of the door;
- a power failure to the door release mechanism or smoke-sensitive devices; or
- the actuation of a fire warning system linked to the door release mechanisms or a fault in that system.

Such fire doors should be labelled with the words 'Automatic Fire Door - Keep Clear'. Where possible, automatic fire doors should be closed at night and have an additional sign to this effect. (Automatic release mechanisms should comply with British Standard 5839: Part 3.)

Other automatic devices are available which operate on different principles - you should consult your local fire authority before installing them.

Emergency escape and fire exit signs

Emergency escape routes and exit doors that are not in common use should be clearly

Part 3 - Further guidance on fire precautions

indicated, as appropriate, by suitable signs. However, in certain circumstances, such as places of public assembly, you should indicate all exit doors. All signs should be in positions where they can be seen clearly. These signs must take the form of a pictogram which may incorporate a directional arrow. The sign can also be supplemented by words such as 'Fire Exit'.

Examples of pictographic fire exit signs

Note: Fire safety signs must comply with the relevant requirements of the Health and Safety (Safety Signs and Signals) Regulations 1996 (see the References section for details of the relevant guidance).

Lighting of escape routes

All escape routes, including external ones, must have sufficient lighting for people to see their way out safely. Emergency escape lighting may be needed if areas of the workplace are without natural daylight or are used at night.

Before providing emergency escape lighting, check the relevant parts of the workplace with the lights off to see whether there is sufficient borrowed light from other sources to illuminate the escape route, eg street lights or unaffected lighting circuits. Where you decide there is insufficient light, you will need to provide some form of emergency lighting. Emergency lighting needs to function not only on the complete failure of the normal lighting, but also on a localised failure if that would present a hazard.

Emergency escape lighting should:

- indicate the escape routes clearly;
- provide illumination along escape routes to allow safe movement towards the final exits; and
- ensure that fire alarm call points and fire-fighting equipment can be readily located.

In addition to emergency escape lighting, it may be necessary to provide other forms of emergency lighting for safety reasons, for example to ensure that manufacturing processes can be shut down safely.

In smaller workplaces and outdoor locations with few people, the emergency escape lighting could take the form of battery-operated torches placed in suitable positions

Part 3 - Further guidance on fire precautions

where employees can quickly get access to them in an emergency, eg on an escape route. In other cases, you should provide an adequate number of electrically operated emergency lighting units, installed to automatically come on if the power to the normal lighting supply circuit, which they are connected to, fails.

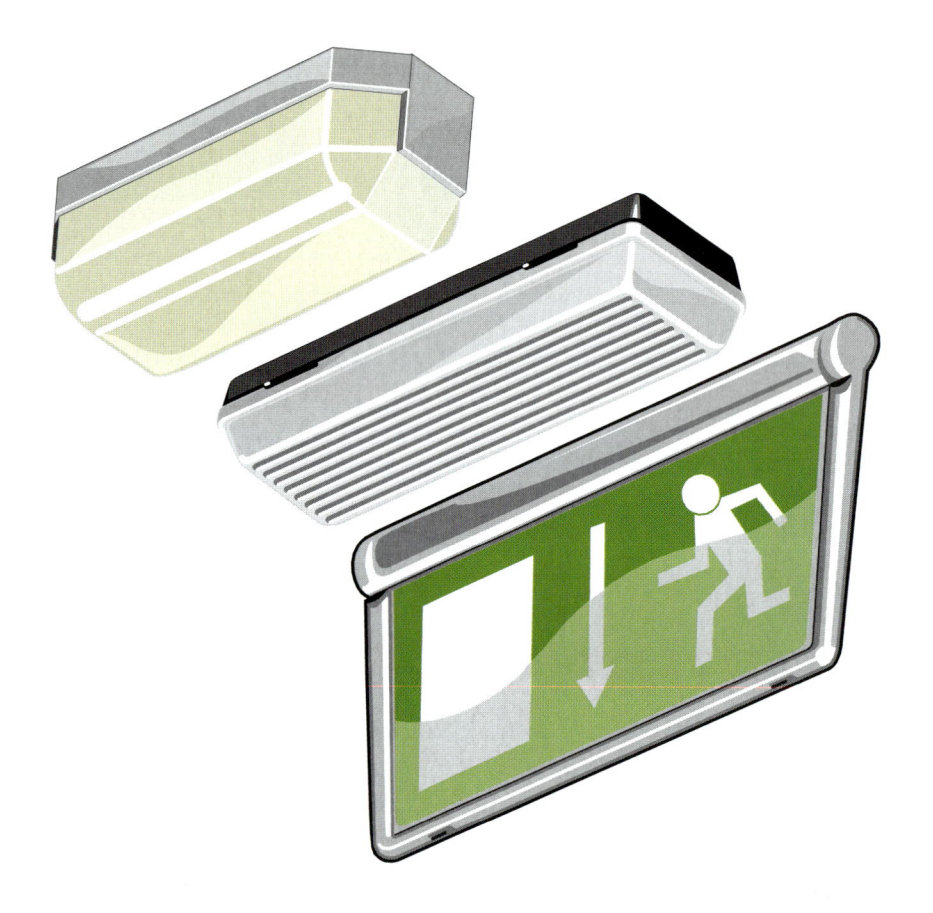

Typical emergency lighting units

Emergency lighting units should be sited to cover specific areas, for example:

- intersections of corridors;
- at each exit door;
- near each staircase so that each flight of stairs receives direct light;
- close to a change in floor level;
- outside each final exit;
- by exit and safety signs that are required elsewhere following the risk assessment;
- within lift cars;
- near fire-fighting equipment; and
- near each fire alarm call point.

Part 3 - Further guidance on fire precautions

The lighting units should be placed as low as possible but at least 2 metres above floor level (measured to the underside of the lighting unit). You may need to consider alternative mounting arrangements in areas where smoke could accumulate and make the lighting ineffective.

Where it is considered that an electrical emergency lighting system is required, the system should be installed in accordance with British Standard 5266: Part 1. You should seek the advice of a competent person who specialises in the installation of these systems.

Smoke control systems for the safety of people

In larger or complex buildings, a smoke control system may be an effective way of keeping means of escape routes clear of smoke so that occupants can evacuate safely.

Smoke exhaust, using either natural smoke exhaust ventilators or powered smoke exhaust ventilators, is arranged so that the hot smoke and gases rise and collect under the ceiling in reservoirs and are then removed through the ventilators. The aim is to keep smoke at a safe height above the heads of people using the escape route, while the fire is still burning.

To achieve this, any smoke and heat exhaust system and its components should be designed and installed by a competent person. Guidance on the design of these systems is in the Building Research Establishment (BRE) reports *Design principles for smoke ventilation in enclosed shopping centres* BR 186, and *Design approaches for smoke control in atrium buildings* BR 258. The British Standard Draft for Development 240 Parts 1 and 2 and the CIBSE Fire Engineering Guide also provide useful guidance. (See the References section for details of these publications.)

Alternatively, a smoke control system using a pressure differential system or depressurisation system can be used to maintain a smoke-free escape route. When using this form of control, the design, installation and ongoing maintenance of the system should be in accordance with British Standard 5588: Part 4.

Buildings under refurbishment

If you decide to refurbish, redecorate or alter the workplace in a way that affects one or more means of escape, you will need to make sure that there are still enough escape routes for the staff (and any others present) to use should a fire occur.

If your workplace has a current fire certificate in force, you must inform your local fire authority before carrying out any structural or material alterations. You may also need Building Regulation approval and should consult the relevant building control authority. Other regulations controlling the safety arrangements in the workplace during construction or alteration may also apply, such as the Construction (Health, Safety and Welfare) Regulations 1996. Your local fire authority will be able to offer you further advice.

Part 3 - Further guidance on fire precautions

FIRE-FIGHTING EQUIPMENT

Portable fire extinguishers

Portable fire extinguishers enable suitably trained people to tackle a fire in its early stages, if they can do so without putting themselves in danger.

All workplaces should be provided with means of fighting fire for use by people in the premises. When you are deciding on the types of extinguisher to provide, you should consider the nature of the materials likely to be found in your workplace.

Fires are classified in accordance with British Standard EN 2 as follows:

Class A   fires involving solid materials where combustion normally takes place with the formation of glowing embers;
Class B   fires involving liquids or liquefiable solids;
Class C   fires involving gases;
Class D   fires involving metals; and
Class F   fires involving cooking oils or fats.

Class A and B fires

Class A fires involve solid materials, usually of organic matter such as wood, paper etc. They can be dealt with using water, foam or multi-purpose powder extinguishers, with water and foam considered the most suitable. Your risk assessment will help you decide how many you need.

Class B fires involve liquids or liquefiable solids such as paints, oils or fats. It would be appropriate to provide extinguishers of foam (including multi-purpose aqueous film-forming foam (AFFF)) carbon dioxide, halon* or dry powder types. Carbon dioxide extinguishers are also suitable for a fire involving electrical equipment.

The fire extinguishers currently available for dealing with Class A or Class B fires should not be used on cooking oil or fat fires (but see 'Class F fires' on page 59).

Class C fires

Dry powder extinguishers may be used on Class C fires. However, you need to consider the circumstances for their use and combine this with action such as stopping the leak, to remove the risk of a subsequent explosion from the build-up of unburnt gas.

Class D fires

None of the extinguishers referred to above will deal effectively with a Class D fire as these involve metals such as aluminium, magnesium, sodium or potassium. Only specially trained personnel using special equipment should tackle such fires. If your

---

*Halon extinguishers are not generally recommended for day-to-day fire risks because of their ozone-depleting potential. They will be banned except for defined essential uses after 31 December 2003. For special risks, one of the other liquefied gas-type extinguishers may be used.

Part 3 - Further guidance on fire precautions

assessment identifies the risk of a fire involving these metals, you should consult your local fire authority about the best way of dealing with it.

Class F fires

Special extinguishers are available for use on fires involving cooking fats and oils, eg chip pans and deep fat friers, but these should only be used by specially trained people. (This is not an established class within the current British Standard but may be added as a new fire classification when the relevant standard is revised. The class is used in this guide for ease of reference.)

Types of portable fire extinguishers

The fire-fighting extinguishing medium in portable extinguishers is expelled by internal pressure, either permanently stored or by means of a gas cartridge. Generally speaking, portable fire extinguishers can be divided into five categories according to the extinguishing medium they contain:

- water;
- foam;
- powder;
- carbon dioxide; and
- vaporising liquids, including halons.

Some fire extinguishers can be used on more than one type of fire. For instance, AFFF extinguishers can be used on both Class 'A' fires and Class 'B' fires. Your fire equipment supplier will be able to advise you.

The most useful form of fire-fighting equipment for general fire risks is the water-type extinguisher or hose reel. One such extinguisher should be provided for approximately each 200 square metres of floorspace, with a minimum of one per floor. If each floor has a hose reel which is known to be in working order and of sufficient length for the floor it serves, there may be no need for water-type extinguishers to be provided.

Areas of special risk involving the use of oil, fats, or electrical equipment may need carbon dioxide, dry powder or other types of extinguisher. If you are not sure what to provide in any given circumstances, your local fire authority will be able to advise you. British Standard 5306: Part 3 provides advice about the selection and allocation of portable fire extinguishers.

Fire extinguishers should conform to a recognised standard such as British Standard EN 3 for new ones and British Standard 5423 for existing ones. For extra assurance, you should look for the British Standard Kitemark, the British Approvals for Fire Equipment (BAFE) mark or the Loss Prevention Council Certification Board (LPCB) mark.

Part 3 - Further guidance on fire precautions

Fire extinguishers may be colour-coded to indicate their type. Previously, the entire body of the extinguisher has been colour-coded, but British Standard EN 3: Part 5 (which came into effect on 1 January 1997) requires that all new fire extinguisher bodies should be red. A zone of colour of up to 5% of the external area, positioned immediately above or within the section used to provide the operating instructions, may be used to identify the type of extinguisher. This zone should be positioned so that it is visible through a horizontal arc of 180° when the extinguisher is correctly mounted. The colour-coding should follow the recommendations of British Standard 7863.

Fire extinguishers, if properly maintained and serviced, may be in service for at least 20 years. So there may be situations where a building will have a mixture of new and old fire extinguishers with the same type of extinguishing medium but with different colour-coded markings. In these cases and to avoid any confusion, it is advisable to ensure that extinguishers of the same type but with different colour-coded markings are not mixed, either at the same location in single-storey buildings or on the same floor level in multi-storey buildings.

Old-style fire extinguishers must not be painted red to try and comply with the new standard, as this would contravene British Standard EN 3 which covers technical changes during the manufacturing stage.

Fire extinguishers should normally be located in conspicuous positions on escape routes, preferably near exit doors. Wherever possible, fire-fighting equipment should be grouped to form fire points. These should be clearly visible or their location clearly and conspicuously indicated so that fire points can be readily identified. Where workplaces are uniform in layout, extinguishers should normally be located at similar positions on each floor.

If for any reason extinguishers are placed in positions hidden from direct view, the Health and Safety (Safety Signs and Signals) Regulations 1996 require that their location should be indicated by signs and, where appropriate, directional arrows. Suitable signs are described in the HSE Guidance on the Regulations (see the References section).

Part 3 - Further guidance on fire precautions

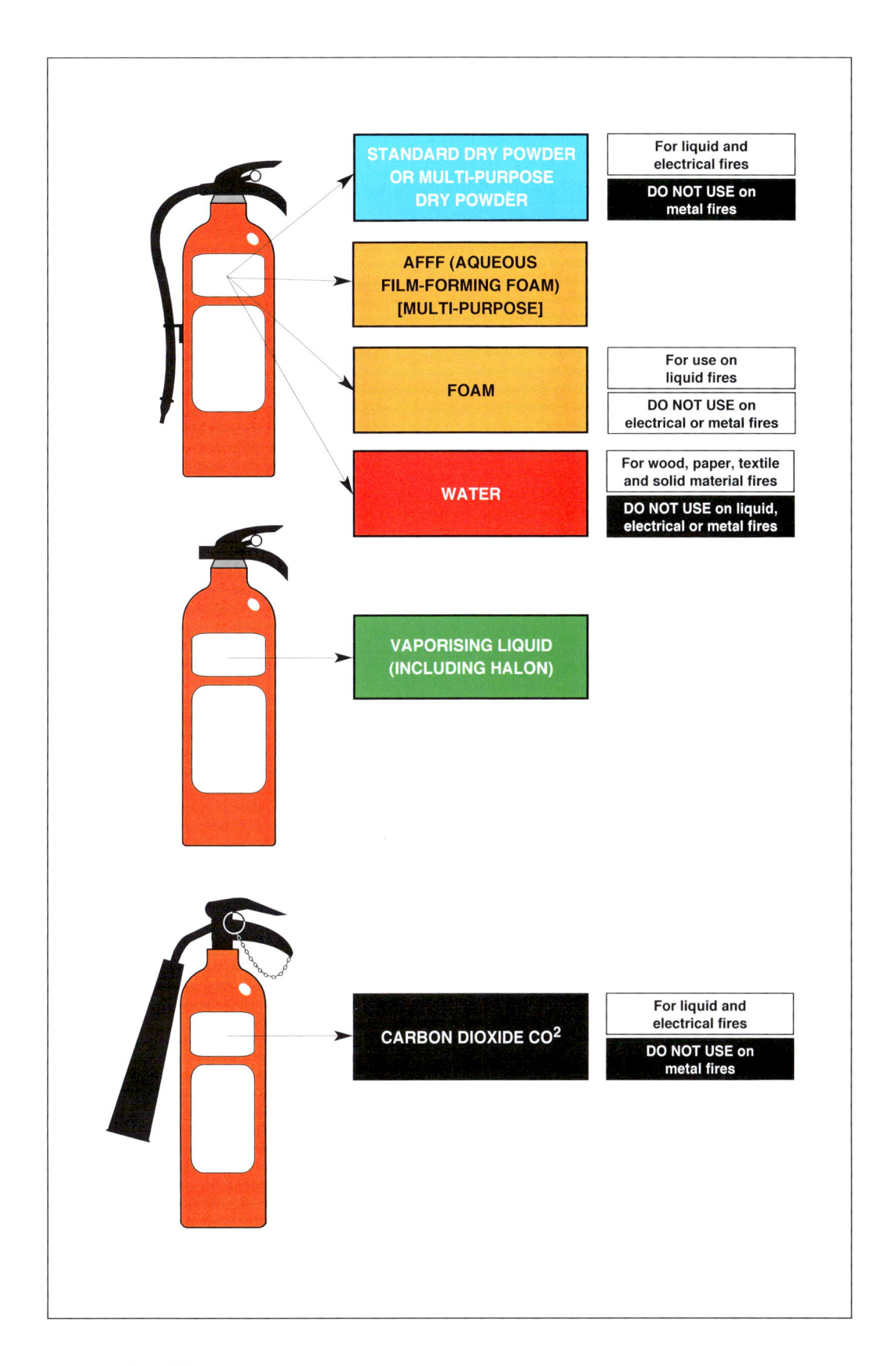

STANDARD DRY POWDER OR MULTI-PURPOSE DRY POWDER	For liquid and electrical fires. DO NOT USE on metal fires
AFFF (AQUEOUS FILM-FORMING FOAM) [MULTI-PURPOSE]	
FOAM	For use on liquid fires. DO NOT USE on electrical or metal fires
WATER	For wood, paper, textile and solid material fires. DO NOT USE on liquid, electrical or metal fires
VAPORISING LIQUID (INCLUDING HALON)	
CARBON DIOXIDE CO_2	For liquid and electrical fires. DO NOT USE on metal fires

Colour-coding of fire extinguishers

Part 3 - Further guidance on fire precautions

Where practicable, fire extinguishers should be securely hung on wall brackets. Where this is impracticable, extinguishers should be placed on a suitable baseplate (not on the floor). To assist in lifting, the carrying handle of larger, heavier extinguishers should be about 1 metre from the floor but smaller, lighter extinguishers may be mounted at a higher level. Make sure that the weight of the equipment falls below the guidelines recommended in the Manual Handling Operations Regulations 1992 (see the References section). This will ensure that extinguishers are easy to handle and use.

Hose reels and fire blankets

Where hose reels are provided, they should be located where they are conspicuous and always accessible, such as in corridors.

Fire blankets should be located in the vicinity of the fire hazard they are to be used on, but in a position that can be safely accessed in the event of a fire. They are classified as either light-duty or heavy-duty. Light-duty fire blankets are suitable for dealing with small fires in containers of cooking oils or fats and fires involving clothing. Heavy-duty fire blankets are for industrial use where there is a need for the blanket to resist penetration by molten materials.

Hose reels and fire blankets should conform with relevant British Standards (see the References section). If you are unsure about the number or type of portable fire-fighting equipment or hose reels you need, you should check with the local fire authority before purchasing any such equipment.

Fixed fire-extinguishing systems

Sprinkler systems

In smaller workplaces, portable fire extinguishers will probably be sufficient to tackle small fires. However, in more complex buildings, or where it is necessary to protect the means of escape and/or the property or contents of the building, it may be necessary to consider a sprinkler system.

Sprinkler systems are traditionally acknowledged as an efficient means of protecting buildings against extensive damage from fire. They are also now acknowledged as an effective means of reducing the risk to life from fire. Systems are being developed which may be suitable for use in workplaces with residential areas, such as care homes and houses in multiple occupancy.

Sprinkler systems need to be specifically designed and installed to the appropriate hazard category in accordance with an approved code of practice (eg British Standard 5306: Part 2 - see the References section, or Technical Bulletins from the Loss Prevention Council - see Further information). This will ensure that that the operation of the system will effectively limit/control the effects of the fire with minimal failures or unwanted operations (these are usually due to inappropriate use or poor maintenance).

Part 3 - Further guidance on fire precautions

Further information on the requirements for sprinkler systems, and their benefits, can be obtained from your local fire authority.

Other fixed fire-fighting systems

In many industrial and commercial premises, fires can pose a serious threat to the safety of employees working in or adjacent to areas involving:

- process machinery;
- electrical switchgear and transformers;
- control- and data-processing equipment; and
- flammable materials storage.

Fires involving these risks can effectively be dealt with by the installation of fixed fire-fighting systems that may be either automatically or manually operated. For example, process equipment and machinery which handles flammable substances (eg printing machines, rolling mills, or oil-filled switchgear), may be protected by extinguishing systems, using dry powder, foam, carbon dioxide or other inert gas. However, recent developments using water mist technology mean that these systems may also be useful, especially in food-processing areas.

Similarly, protection of control- and data-processing equipment may be achieved by systems designed to totally fill the room or the cabinets containing such equipment with a gas-flooding extinguishing medium to a specified concentration. These types of systems use a range of gas extinguishing media. Where there is a possibility that these may discharge into occupied areas, you need to ensure that the resulting concentration of the extinguishing medium will not be harmful to anybody present.

Where necessary, protection of large-scale storage facilities of flammable materials, especially in bulk tanks, may also be achieved by fixed deluge water or foam systems.

The design and installation of fixed fire-fighting systems requires a high level of expertise, including the ability to carry out a thorough risk assessment and select the appropriate system and fire-fighting medium. Such systems have to be specially designed, and can be expensive. If you are considering installing such a system, you should liaise with the relevant enforcing authority and consult a reputable company at an early stage. The British Fire Protection Systems Association can supply you with a list of companies in your area that undertake this type of work (see the 'Further information' section for details).

Part 3 - Further guidance on fire precautions

INSTALLATION, MAINTENANCE AND TESTING OF FIRE PRECAUTIONS AND EQUIPMENT

Part 2 of this guide highlights the items that should be maintained and tested to ensure that the safety features, put in place following the risk assessment, are in the right location and function correctly when required. This section looks at what should be done to ensure that equipment and systems will be effective when needed.

Any electrical fire detection and fire warning systems must be kept switched on when the workplace is in use. This also applies to systems which are remotely monitored, when the workplace is unoccupied. In other cases, it is advisable to keep them turned on, even when the workplace is not in use.

All equipment provided to safeguard the safety of employees in the workplace, eg fire doors and fire-fighting equipment, should be regularly checked and maintained by a competent person in accordance with the relevant British Standard (see the References section) and the manufacturer's recommendations.

It is advisable to keep a record of any maintenance and testing of equipment, as this will be useful if you are asked by the fire authority to demonstrate that you have an effective system in place. (You must do this if your premises are covered by a fire certificate.)

It is also good practice for employers and employees to carry out routine checks on a daily basis. This would include checking that:

- the control panel shows that all electrical fire detection and alarm systems are operating normally, or ensuring that any faults indicated are recorded and dealt with;
- all emergency lighting systems that include signs are lit and any defects recorded and dealt with;
- all escape routes, including passageways, corridors, stairways and external routes, are clear of obstruction, free of slipping and tripping hazards and available for use when the premises are occupied;
- all fastenings on doors along escape routes operate freely, so that they can be opened quickly in an emergency without delay;
- all self-closing devices and automatic door holders/releases work correctly, and that each door closes completely (this check should include a look at any fitted flexible edge seals, to ensure that they can still provide an effective smoke seal);
- all exit and directional signs are checked to make sure that they are correctly positioned and can be clearly seen at all times; and
- all fire extinguishers are in position, have not been discharged, are at the correct pressure and have not suffered any obvious damage.

Any defects should be reported in line with your own procedures and repaired as soon as possible.

Part 3 - Further guidance on fire precautions

Maintenance and service schedules for fire protection equipment should comply with the relevant British Standard (see the References section) and the manufacturer's instructions. The following paragraphs give an indication of the levels of service required.

Fire detection and fire warning systems

All manually operated fire alarms, such as rotary gongs, should be tested weekly to ensure that they work and can still be heard throughout the workplace.

Electrical fire detection and fire warning systems should be tested weekly for function and to check whether they can be heard throughout the area covered. Make sure they can be seen or heard, particularly by disabled people, and that voice alarms can be understood. They should also be inspected and tested, quarterly and annually, by a competent person.

Self-contained, domestic-type smoke alarms should be tested weekly and cleaned annually. Replaceable batteries should be changed at least once a year (except for extended life batteries where the manufacturer's recommendations should be followed) or when the low battery warning device operates.

Fire extinguishers and hose reels

Basic inspection procedures for portable extinguishers should be carried out on a weekly basis and should include:

- checking the safety clip and indicating devices to determine whether the extinguisher has been operated; and
- checking the extinguisher for any external corrosion, dents or other damage that could impair the safe operation of the extinguisher.

As well as this, servicing should be carried out by a competent person as follows:

- basic annual service;
- extended service every five years (see table on page 66); and
- overhaul/replacement every 20 years.

Hose reels should be checked weekly to make sure that they are not damaged or obstructed. They should be serviced annually by a competent person.

Part 3 - Further guidance on fire precautions

	EXTENDED SERVICE PROCEDURES
	Inspection requirement
1	Check the functioning of the pressure-indicating devices, where fitted, of stored pressure portable, fire portable extinguishers, according to the instructions of the portable fire extinguisher supplier and/or holder of the approval.
2	Perform the test discharge or empty all portable fire extinguishers, except the halon type.
3	Examine the extinguishing media according to the instructions of the portable fire extinguisher supplier and/or holder of the approval.
4	Examine in detail for corrosion, damage, dents, gouges: - head cap and valves; - indicators; and - discharge hose and nozzle.
5	Examine the body internally in detail for corrosion, dents, cuts, gouges or lining damage. Pay special attention to the welds. If you are in doubt about welds, follow the instructions of the portable fire extinguisher supplier and/or holder of the approval.
6	Examine and check all closures for thread wear, damage and coating as applicable.
7	Return to operational condition. Reassemble the portable fire extinguisher according to the instructions of the portable fire extinguisher supplier and/or holder of the approval and charge.

Table of extended service procedure to be used by a competent person

Fixed fire-fighting systems

Where automatic sprinklers or other fixed fire-fighting systems are installed, they should be tested in accordance with the manufacturer's/installer's specifications (this test does not mean activating the sprinkler head(s) to see if water flows from the system). They should also be serviced annually by a competent person.

Systems employing high-pressure gas storage cylinders, for example carbon dioxide-based systems, should be maintained and inspected by a competent person.

Part 3 - Further guidance on fire precautions

Portable lamps, torches and radios

Where portable items such as lamps, torches and radios are to be used as part of the fire precautions for the workplace, it is important to ensure that they are appropriate and suitable for the purpose. Additionally, if they may be used in a potentially explosive atmosphere, they should comply with appropriate standards and certification.

A named individual should be nominated as responsible for the equipment, and you should have a system in place to check that the equipment is available and operates correctly. This check should be carried out daily, or at the beginning of each shift. You should also operate a fault-recording and repair system. A sufficient number of spare torches and radios and their batteries should be readily available to replace any that are found to be faulty.

Emergency lighting

Automatic emergency escape lighting equipment should be inspected by a competent person monthly, six-monthly and three-yearly, in accordance with the schedules set out in the relevant British Standard and the manufacturer's recommendations.

Smoke control systems

Where the design of the building incorporates smoke control systems to protect life, the system should be maintained in accordance with the manufacturer's instructions or the relevant British Standard.

Quality assurance of fire protection equipment

Fire protection products and related services should be fit for their purpose and properly installed and maintained in accordance with the manufacturer's instructions or the relevant British Standard.

Third-party certification schemes for fire protection products and related services are an effective means of providing the fullest possible assurances, offering a level of quality, reliability and safety that non-certificated products may lack. This does not mean goods and services that are not third-party-approved are less reliable, but there is no obvious way in which this can be demonstrated.

Third-party quality assurance can offer great comfort to employers, both as a means of satisfying you that the goods and services you have purchased are fit for purpose, and as a means of demonstrating that you have complied with the law.

Your local fire authority can provide further details about independent third-party quality assurance schemes and the various organisations that administer them.

Part 3 - Further guidance on fire precautions

HISTORIC AND LISTED BUILDINGS

Fires in historic buildings that are workplaces not only carry a risk of loss of life and earnings but they can also mean the loss of an irreplaceable part of our heritage. Because these historic buildings are so valuable, any proposed changes, including fire precautions etc, must be carefully considered and carried out with the intention of 'minimum intervention' in the building's fabric.

Most building works are subject to building control under Building Regulations. Historic and listed buildings are also subject to controls under planning legislation. In the latter case, any proposed building work may therefore also require listed building consent from the planning authority. Such controls will apply in the case of any work which could affect the character of the building, such as the alteration of doors or door fittings to increase their fire-resistance, the provision of new fire-resisting doors or the treatment of panelling and internal woodwork etc.

All applications for consent to carry out building, alteration or demolition work on Grade I or Grade II listed buildings (Category A or B in Scotland) will be notified by the planning authority to the Secretary of State (or Scottish Ministers). Consent may only be granted by the planning authority if the Secretary of State indicates that a personal determination by the Secretary of State is not necessary in the circumstances of the case. You should seek advice from your local building control authority or other building approvals body at an early stage if any building works are proposed.

It is important to be flexible in assessing the fire safety measures that will be appropriate for buildings in these categories, particularly when you need to ensure that structural matters are in character with the rest of the building. It is also important to ensure that the work does not cause unacceptable damage to the fabric of the building.

Where a fire certificate or some other type of fire safety approval is required (see Annex A) it is important that you advise the enforcing authority of all the important facts, including the Historic or Listed Building status.

If there are substantial practical difficulties in upgrading the building to an acceptable standard of fire safety in the conventional way, fire safety engineering may provide an acceptable alternative. Before considering such a solution, you should check with the local building control authority or other building approvals body whether this approach is acceptable under the building legislation which applies to your workplace.

A fire safety engineering approach that takes the total fire safety package into account can provide a more fundamental and economic solution than more prescriptive approaches to fire safety.

In some instances and particularly where members of the public are admitted, if an adequate fire safety solution cannot be achieved without unacceptable alteration to the fabric or character of the building, there are two options:

Part 3 - Further guidance on fire precautions

- limit the number of occupants in the workplace; or
- stop using part of the workplace for that particular purpose.

However, an increase in supervisory employees and effective surveillance and supervision of evacuation procedures may, in some circumstances, compensate for shortcomings in some structural features.

Details of specific advice about fire precautions in historic buildings are given in the References section.

DISABLED PEOPLE

Legislation dealing with the needs of disabled people does not make any specific requirements regarding means of escape in case of fire. However, the Disability Discrimination Act 1995 requires employers to make 'reasonable adjustments' to their premises to ensure that no employee is at a disadvantage. This includes ensuring that disabled people can leave the premises safely in the event of fire.

As an employer, you are therefore under an obligation to ensure that your emergency plan takes account of disabled people. It is essential that you identify the special needs of any disabled employees when planning your fire safety arrangements and evacuation procedures. You will also need to consider other less able-bodied people who may have access to the workplace.

You may have to take account of the difficulties people with a wide range of physical and/or mental impairment can have in getting in and out of the workplace (particularly in an emergency).

If any of your employees have disabilities, your emergency plan should be developed in conjunction with them, taking their disabilities into account.

Means of escape

Means of escape for disabled people in new or altered buildings is provided for by building regulations and, in existing buildings, by fire safety legislation (eg the Fire Regulations and the Fire Precautions Act etc, see Annex A).

British Standard 5588: Part 8 gives detailed guidance regarding most new or altered buildings (see the References section). The code should also be followed wherever possible in relation to existing buildings. However, it is important to note that the relevant legislation has to be complied with in the event of any conflict with the code. The following guidance is based upon some of the recommendations in the British Standard but the code itself should be referred to for greater detail.

Part 3 - Further guidance on fire precautions

Use of lifts as means of escape

Unlike normal passenger lifts, it is essential that a lift which is to be used to evacuate disabled people can continue to be operated with a reasonable degree of safety when there is a fire in the building.

Although it is not necessary to provide a lift specifically for the evacuation of disabled people, a fire-fighting lift (see British Standard 5588: Part 5), which is provided principally for the use of the fire service, may be used to evacuate disabled people before the fire brigade arrive. Another acceptable way of evacuating disabled people requiring assistance is a passenger (evacuation) lift (see British Standards 5810 and 5655).

Normally, only disabled people should rely on a lift as a means of escape and only then if it is an evacuation lift specially designed for the evacuation of disabled people as described in British Standard 5588: Part 8. It must be under the control of the management using an agreed evacuation procedure. The lift should be provided with a means of switching control from general use to the car itself, so that an operator can take it to those floors from which disabled people need to be evacuated.

Refuges

Because of the limits on distances to be travelled for means of escape, most disabled people should be able to reach the safety of a protected escape route or final exit independently. However, some disabled people, for example those who rely upon a wheelchair, will not be able to use stairways without assistance. For these people it may be necessary to provide refuges on all storeys other than in those small buildings of limited height (eg where the distance of travel to a final exit is so limited that refuges are unnecessary). You should check with your local fire authority before considering providing refuges.

In this situation, a refuge is an area that is both separated from the fire by fire-resisting construction and which has access via a safe route to a storey exit. It provides a temporarily safe space for disabled people to wait for others to help them evacuate.

Examples of satisfactory refuges include:

- an enclosure such as a compartment, protected lobby, protected corridor or protected stairway (see 'Technical terms relating to means of escape' earlier in Part 3);
- an area in the open air such as a flat roof, balcony, podium or similar space which is sufficiently protected (or remote) from any fire risk and provided with its own means of escape; and
- any other arrangements which satisfy the general principles outlined above and which provide at least an equal measure of safety.

The refuge needs to be big enough to allow wheelchair use and to allow the user to manoeuvre into the wheelchair space without undue difficulty. It is essential that the

Part 3 - Further guidance on fire precautions

location of any wheelchair spaces within refuges does not adversely affect the means of escape for other people.

Ageing

Older people would generally benefit from facilities provided for people with a disability in public buildings but not all are in need of them. Only a minority of elderly people would be classified as having a disability. It is a mistake to equate old age with physical disability, but the age of the likely occupants will need to be considered in any calculations for means of escape facilities.

Assisting the less able-bodied

If people use a wheelchair, or can only move about with the use of walking aids, their disability is obvious. But disabilities can sometimes be less obvious than this and staff should be vigilant in an emergency, so that help can be given to those members of the public who need it most, including the very young and the elderly. If members of staff have disabilities, the emergency plan should be developed in conjunction with them, taking this into account.

Assisting wheelchair users and people with impaired mobility

In drawing up an evacuation plan, you should consider how wheelchair users and people with impaired mobility can be assisted. Some types of lift may be used but, where stairs need to be negotiated and people with disabilities may have to be carried, you should consider training enough able-bodied members of staff in the correct methods of doing so.

With a number of individuals, their impaired mobility may only be temporary. Members of staff in the advanced stages of pregnancy or with broken limbs will only be temporarily affected, but you must consider their special needs in your emergency plan.

Assisting people with impaired vision

People with impaired vision or colour perception may experience difficulty in seeing or recognising fire safety signs. However, many people are able to read print if it is sufficiently large and well designed with a good, clear typeface. Signs should therefore be designed and sited so that they can be seen easily and are readily distinguishable.

Good lighting and the use of simple colour contrasts can also help visually impaired people find their way around. If you need advice about this, you can contact the Royal National Institute for the Blind or the National Federation of the Blind of the United Kingdom (see Further information section).

Staff with impaired vision should be familiarised with escape routes, especially those which are not in general use. In an evacuation of a building, a sighted person should lead

Part 3 - Further guidance on fire precautions

such members of staff to safety. Similar assistance should be offered to guide dog owners, with the owner retaining control of the dog. A normally sighted person should remain with staff with impaired vision until the emergency is over.

In the evacuation of the premises, it is recommended that a sighted person should lead, inviting the other person to grasp their elbow, as this will enable the person being assisted to walk half a step behind and thereby gain information about doors and steps etc. Similar assistance should be offered to guide dog owners, with the owner retaining control of the dog.

Employees need to be clear what to do if the guide dog remains in the building and refuses to leave. Human life should not be put at risk if the dog refuses to leave.

Assisting people with impaired hearing

Although people with impaired hearing may experience difficulty in hearing a fire alarm, they may not be completely insensitive to sound; some may be able to hear a conventional alarm signal and require no special provision. However, where a member of staff or the public is known to have difficulty, someone should be given the responsibility of alerting the individual concerned. You will need to have cover for leave and other absences. You can also get advice from the Royal National Institute for Deaf People (see Further information).

You should consult your workforce before and after the installation of alternative alarm signals because of possible unwanted side effects and to ensure that the system is effective. Induction loop systems used in some premises for audio communication with people using suitable hearing aids are not acceptable as a means of alerting people with impaired hearing in the event of fire. However, if such systems are in normal use in your workplace, they may be used to supplement the alarm.

Assisting people with learning difficulties or mental illness

Any staff with learning difficulties or mental illness must be told what they should do in the event of fire. Arrangements should be made to ensure that they are assisted and reassured in a fire situation and are accompanied to a place of safety; they should not be left unattended. Advice may also be sought from MENCAP (see Further information) or from local residential or day services for people with learning difficulties.

Sources of advice

The names and addresses of organisations representing people with disabilities and sensory-impaired people can be found in the Further information section. Details of similar organisations can be found in the Yellow Pages.

ANNEXES

ANNEX A - OTHER LEGISLATION THAT MAY APPLY TO YOUR WORKPLACE

Premises which require a Fire Certificate under the Fire Precautions Act 1971 (or the Fire Services (Northern Ireland) Order 1984 as amended)

The use of certain types of premises has been designated by the Secretary of State as requiring a fire certificate under the Fire Precautions Act 1971 (in Northern Ireland under the Fire Services (Northern Ireland) Order 1984 as amended). There are two designating orders in force in Great Britain (and four in Northern Ireland). One relates to larger hotels and boarding houses and the other to those factories, offices, shops and railway premises in which people are employed to work.

The first designating order (the Fire Precautions (Hotels and Boarding Houses) Order 1972) requires a fire certificate when premises are used as a hotel or boarding house which will provide sleeping accommodation for more than six people (whether employees or guests) or if they provide sleeping accommodation for employees or guests elsewhere than on the ground or first floors of the premises. (In Northern Ireland, this requirement is under the Fire Services (Hotels and Boarding Houses) Order 1985 as amended.)

The second designating order (the Fire Precautions (Factories, Offices, Shops and Railway Premises) Order 1989) (in Northern Ireland, this requirement is under the Fire Services (Factory, Office and Shop Premises) Order (Northern Ireland) 1993) requires that a fire certificate must be applied for when more than 20 people are at work at any one time in your workplace, or more than 10 are at work at any one time elsewhere than on the ground floor.

In buildings in multiple occupation containing two or more similar premises, a certificate must be sought when the number of workers exceeds the above totals. Fire certificates are also required for factory premises where explosive or highly flammable materials are stored or used, regardless of the number of people at work, unless the fire authority has determined otherwise. (The fire authority may exempt premises from the certification requirement if they consider them to be of low risk.)

In Northern Ireland, in addition to the above, leisure premises and betting, gaming and amusement premises have also been designated under the Fire Services (Northern Ireland) Order 1984 as amended:

- The Fire Services (Leisure Premises) Order (Northern Ireland) 1985 requires (with some exceptions) a fire certificate in respect of premises used as recreational facilities by a district council under Article 9(1)(a) of the Recreation and Youth Services (Northern Ireland) Order 1973 or used for entertainment, recreation or instruction by a university or college.
- The Fire Services (Betting, Gaming, and Amusement Premises) Order (Northern Ireland) 1987 requires a fire certificate for premises for which a bookmaking licence, a track betting licence, a bingo club licence, an amusement permit or a leisure permit is required under the Betting, Gaming, Lotteries and Amusements (Northern Ireland) Order 1985.

Annex A

To apply for a fire certificate you should ask the fire safety office of your local fire authority for an application form. When it is completed, you should return the form.

In many cases where the requirements of the Fire Regulations are complied with, this will provide sufficient protection from fire for the fire authority to issue a fire certificate without any further action being needed. This is likely to be the case if you have taken full account of the other people who may be present and the means of escape in case of fire.

In cases where both the Fire Regulations and the requirement for a fire certificate apply, it is advisable to discuss the fire precautions you propose, as a result of your risk assessment, with the fire authority before putting the precautions in place. This will allow any special requirements, which may be needed for the fire certificate to be issued, to be considered at the same time and help to avoid any unnecessary expenditure.

If you already hold a fire certificate under the Fire Precautions Act 1971 (or the Fire Services (Northern Ireland) Order 1984 (as amended)) the law requires you to notify the fire authority before making any change to the fire precautions in your workplace as a result of your risk assessment if such change affects the terms and conditions of your fire certificate.

The Fire Certificates (Special Premises) Regulations 1976

These Regulations define certain premises that will require a fire certificate, based on the storage or use of quantities of hazardous substances above specified threshold quantities, or based on particular hazardous activities. Details are given in Schedule 1 to the Regulations - advice can be sought from the Health and Safety Executive (HSE).

The Regulations apply to the whole site and not to individual buildings or plant (with the exception of licensed explosive factories). Most premises which fall within these Regulations have already been certificated and the application of these Regulations will be rare for new premises. An application for a fire certificate under the special premises regulations should be made to HSE when the use or storage of a relevant substance reaches the specified limit.

The Explosives Act 1875

The manufacture and storage of explosives are regulated under the Explosives Act 1875. The Act requires that all manufacturers of explosives must be licensed by HSE, as must the largest explosives stores; smaller stores must either have a licence from the local authority or be registered with it. The licence conditions normally contain fire safety measures.

Workplaces which require a licence

A number of uses of workplaces require a licence from the local authority or licensing magistrate. These licences can impose additional fire safety requirements which may go

Annex A

beyond the minimum levels needed by the Fire Regulations. The most common uses of workplaces which are subject to licensing control are those which involve:

- the sale of alcohol;
- music and dancing;
- theatrical performances;
- the showing of films;
- gambling;
- sporting activities; and
- other forms of public entertainment.

If your workplace is, or may be, subject to licensing control, it is advisable to discuss the findings of your risk assessment with the fire authority before putting your proposals for fire safety measures into place. This can help you avoid unnecessary expenditure.

If you already have a licence, you should discuss any proposals you may have for changes to the fire precautions with the fire authority before approaching the authority who issued your licence.

Registration schemes

Some uses of premises are required to be registered with the local authority or other registrar. These uses can include:

- nursing homes;
- residential care homes;
- children's homes; and
- independent schools.

The requirements of registration schemes usually contain fire safety provisions and changes to the fire precautions will often need the agreement of the registering authority. If your workplace is, or may be, subject to such a scheme, it is advisable to discuss the findings of your risk assessment with the fire authority before putting your proposals for fire safety measures into place. This can help you avoid unnecessary expenditure.

If you are already registered, you should discuss any proposals you may have for changes to the fire precautions with the fire authority before approaching the registration authority about your proposals for change.

Building Regulations

In England and Wales the Building Regulations 1991 (in Scotland the Building Standards (Scotland) Regulations 1991, in Northern Ireland the Building Regulations (Northern Ireland) 1994) apply to new buildings and to building work such as the erection, extension or material alteration of an existing building. They also apply where there is a material change of use.

Annex A

The Regulations impose fire safety requirements covering matters such as:

- means of escape in case of fire;
- structural stability;
- fire-resistance of elements and structure;
- compartmentation to inhibit fire spread;
- reduction of spread of flame over surfaces of walls and ceilings;
- space separation between buildings to reduce the risk of fire spread from one building to another; and
- access for fire appliances and assistance to the fire brigade.

The standard of provision is related to the size and height of the building and the use to which it is put. In Scotland the Building Standards (Scotland) Regulations 1991 contain different requirements for the storage of materials that give rise to fire hazards. Where it is proposed to erect a new building, to carry out building work or to make a material change of use, application should be made to your local building control authority or other building approval body.

Other legislation that may apply

If you believe other legislation with fire safety provisions (such as sports grounds safety legislation) may apply to your workplace or are simply unsure what legislation applies, you should contact the fire safety office of your local fire authority. They will be able to tell you which provisions apply in the particular circumstances of your workplace and who you should contact about those provisions.

Health and Safety at Work etc Act 1974

This Act is concerned with the health, safety and welfare of people at work, and with protecting those who are not at work (members of the public etc) from risks to their health and safety arising from work activities. The Act and its relevant statutory provisions cover the risk of fire.

Management of Health and Safety at Work etc Regulations 1992 (as amended)

These Regulations require employers and the self-employed to assess the risks to workers and others who may be affected by their undertakings, so that they can decide what measures need to be taken to comply with health and safety law.

The Regulations require you to implement appropriate arrangements for managing health and safety. Health surveillance (where appropriate), emergency planning, and the provision of information and training are also included. There is an Approved Code of Practice on these Regulations (see the References section).

Annex B

ANNEX B - ENFORCEMENT OF THE FIRE REGULATIONS

In most workplaces, the local fire authority enforces the Fire Regulations and the parts of the Management of Health and Safety at Work Regulations 1992 (MHSW Regulations) dealing with general precautions for the safety of people in case of fire.

Other requirements for preventing fires occurring and taking measures to reduce their severity, under the Health and Safety at Work etc Act 1974 and MHSW Regulations, are enforced by the Health and Safety Executive or the local authority, depending on the activity in the premises.

Special arrangements apply to workplaces owned or occupied by the Crown. For these workplaces, HM Inspectors of Fire Services of the Crown Premises Inspection Groups are responsible for the application of the Fire Regulations and their enforcement. (In Northern Ireland the Department of Economic Development has this responsibility.) Ministry of Defence premises are dealt with by the MOD's Defence Fire Services.

For workplaces which are subject to the Fire Regulations, if the fire authority considers that any provision of the Regulations has not been complied with in respect of your workplace or the employees who work there, they can serve a notice requiring you to improve your fire precautions. Such a notice is known as an enforcement notice and failure to comply with it is a criminal offence. You must be given a reasonable amount of time to comply and can appeal (within 21 days) against the notice to a magistrates' court (in Scotland, the Sheriff Court). If an appeal is lodged the notice will be held in abeyance until the courts have heard the appeal and either upheld the notice, cancelled it or amended it.

In workplaces used by more than one employer, a notice may also be served on any other person who has control, to any extent, over parts of the workplace. The same appeal provisions for employers also apply for these people.

In very serious cases, which are a serious threat to life, the fire authority can serve you with a notice under section 10 of the Fire Precautions Act 1971 (or the Northern Ireland equivalent). This notice can prohibit or restrict the use of your workplace until the risk to your employees or other people has been reduced; failure to comply with it is a criminal offence. You can appeal against the notice to a magistrates' court (in Scotland, the Sheriff Court) but it will remain in force until such time as the court say otherwise.

A failure to comply with the Fire Regulations which places one or more employees at serious risk in case of fire is, in itself, a criminal offence. In any proceedings for such an offence, it is a defence for people charged to prove that they took all reasonable precautions and exercised all due diligence to avoid the commission of the offence.

An employer, or other person, found guilty by a court of either failing to comply with the terms of an enforcement notice, or of placing employees at serious risk by failing to comply with the Fire Regulations, may be sentenced:

- on summary conviction, to a fine not exceeding the statutory maximum; or
- on indictment, to a fine or up to two years' imprisonment, or both.

REFERENCES

HOW TO OBTAIN PUBLICATIONS

British Standards

British Standards are available from
BSI Sales and Customer Services,
389 Chiswick High Road,
London W4 4AL.
Tel: 0208 996 7000
Fax: 0208 996 7001

HSE publications

HSE priced and free publications are available by mail order from
HSE Books, PO Box 1999, Sudbury, Suffolk CO10 2WA.
Tel: 01787 881165
Fax: 01787 313995
HSE priced publications are also available from good booksellers.

Stationery Office (previously HMSO) publications

Stationery Office publications are available from
The Publications Centre, PO Box 276,
London SW8 5DT.
Tel: 0870 600 5522
Fax: 0870 600 5533
They are also available from bookshops.

Arson Prevention Bureau publications

Arson Prevention Bureau publications are available from
51 Gresham Street, London EC2V 7HQ.
Tel: 0207 216 7474

British Standards

BS476: *Fire tests on building materials and structures*

Part 7: *Method for classification of the surface spread of flame of products*

Part 22: *Methods for determination of the fire-resistance of non-loadbearing elements of construction*

BS 5266: *Emergency lighting*

Part 1: *Code of practice for the emergency lighting of premises other than cinemas and certain other specified premises used for entertainment*

BS 5306 *Fire-extinguishing installations and equipment on premises*

Part 1: *Hydrant systems, hose reels and foam inlets*

Part 2: *Specification for sprinkler systems*

Part 3: *Code of practice for selection, installation and maintenance of portable fire extinguishers*

Part 4: *Specification for carbon dioxide systems*

BS 5423: *Portable fire extinguishers*

BS 5446 *Components of automatic fire alarm systems for residential premises*

Part 1: *Specifications for self-contained smoke alarms and point-type smoke detectors*

BS 5499: *Fire safety signs, notices and graphic symbols*

BS 5502: *Buildings and structures for agriculture*

Part 23: *Buildings and structures for agriculture: Code of practice for fire precautions*

BS 5588: 1997 *Fire precautions in the design, construction and use of buildings*

Part 2: *Code of practice for shops*

Part 4: *Code of practice for smoke control in protected escape routes using pressure differential*

Part 5: *Code of practice for fire-fighting stairs and lifts*

Part 6: *Code of practice for places of assembly*

References

Part 7: *Code of practice for the incorporation of atria in buildings*

Part 8: *Code of practice for means of escape for disabled people*

Part 9: *Code of practice for ventilation and air conditioning ductwork*

Part 10: *Code of practice for shopping complexes*

Part 11: *Code of practice for shops, offices, industrial storage and other similar buildings*

BS 5655: *Lifts and service lifts*

BS 5725 *Emergency exit devices*

BS 5810: *Code of practice for access for the disabled to buildings*

BS 5839: *Fire detection and alarm systems for buildings*

Part 1: *Code of practice for the design, installation and servicing of fire detection and alarm systems for buildings*

Part 3: *Specification for automatic release mechanisms for certain fire protection equipment*

Part 6: *Code of practice for the design, installation and servicing of fire detection and alarm systems in dwellings*

Part 8: *Code of practice for the design, installation and servicing of voice alarm systems*

BS 5908: *Code of practice for fire precautions in the chemical and allied industries*

BS 7671: *Requirements for electrical installation: IEE Wiring Regulations*

BS 7863: *Recommendations for colour coding to indicate the extinguishing media contained in portable fire extinguishers*

BS EN 3: *Portable fire extinguishers*

Part 5: *Portable fire extinguishers - specifications and supplementary tests*

BS EN 2: *Classification of fires*

British Standard for Development 240: *Fire safety engineering in buildings*

Part 1: *Guide to the application of fire safety engineering principles*

Part 2: *Commentary on the equations given in Part 1*

HSE publications

A guide to risk assessment requirements INDG218 HSE Books 1996 Single copies free, multiple copies in priced packs ISBN 0 7176 1211 2

A guide to the Health and Safety (Consultation with Employees) Regulations 1996 L95 HSE Books 1996 ISBN 0 7176 1234 1

Assessment of fire hazards from solid materials and the precautions required for their safe storage and use: A guide for manufacturers, suppliers, storekeepers and users HSE Books 1991 ISBN 0 11 885654 5

Buying new machinery INDG271 HSE Books 1998 Single copies free, multiple copies in priced packs ISBN 0 7176 1559 6

Chemical warehousing: The storage of packaged dangerous substances HSG71 HSE Books 1998 ISBN 0 7176 1484 0

Dispensing petrol: Assessing and controlling the risk of fire and explosion at sites where petrol is stored and dispensed as a fuel HSG146 1996 HSE Books 1996 ISBN 0 7176 1048 9

Electricity on the farm Information Sheet AS17 HSE Books 1990

Dust explosions in the food industry Information Sheet FIS2 HSE Books 1993

Fairgrounds and amusement parks: Guidance on safe practice HSG175 HSE Books 1997 ISBN 0 7176 1174 4

Fire safety in construction: Guidance for clients, designers and those managing and carrying out construction work involving significant fire risks HSG168 HSE Books 1997 ISBN 0 7176 1332 1

Fire safety in the paper and board industry HSE Books 1995 ISBN 0 7176 0841 7

Fire safety in the printing industry HSE Books 1992 ISBN 0 11 886375 4

Five steps to risk assessment INDG163 HSE Books 1998 Single copies free, multiple copies in priced packs ISBN 0 7176 1565 0

References

Flame arresters: Preventing the spread of fires and explosions in equipment that contains flammable gases and vapours HSG158 HSE Books 1996 ISBN 0 7176 1191 4

Lift trucks in potentially flammable atmospheres HSG113 HSE Books 1996 ISBN 0 7176 0706 2

Maintaining portable and transportable electrical equipment HSG107 HSE Books 1994 ISBN 0 7176 0715 1

Maintaining portable electrical equipment in hotels and tourist accommodation INDG237 HSE Books 1996 Single copies free, multiple copies in priced packs ISBN 0 7176 1273 2

Maintaining portable electrical equipment in offices and other low risk environments INDG236 HSE Books 1996 Single copies free, multiple copies in priced packs ISBN 0 7176 1272 4

Management of health and safety at work. Management of Health and Safety at Work Regulations 1992. Approved Code of Practice L21 HSE Books 1992 ISBN 0 7176 0412 8

Manual handling. Manual Handling Operations Regulations 1992. Guidance on Regulations L23 HSE Books 1998 ISBN 0 7176 2415 3

Memorandum on the Electricity at Work Regulations 1989. Guidance on Regulations HSR25 HSE Books 1989 ISBN 0 7176 1602 9

Take care with oxygen: fire and explosion hazards in the use of oxygen HSE8 (rev1) HSE Books 1999

Permit to work systems INDG98 HSE Books 1997 Single copies free, multiple copies in priced packs ISBN 07176 1331 3

Petrol filling stations: Construction and operation HSG41 HSE Books 1990 ISBN 0 7176 0461 6

Safe handling of combustible dusts: Precautions against explosions HSG103 HSE Books 1991 ISBN 0 7176 0725 9

Safety in the installation and use of gas systems and appliances. The Gas Safety (Installations and Use) Regulations 1998. Approved Code of Practice and Guidance HSE Books1998 ISBN 0 7176 1635 5

Safety signs and signals. Health and Safety (Safety Signs and Signals) Regulations 1996 Guidance on Regulations L64 HSE Books 1997 ISBN 0 7176 0870 0

Safe use and handling of flammable liquids HSG140 HSE Books 1996 ISBN 0 7176 0967 7

Safe use and storage of cellular plastics HSG92 HSE Books 1996 ISBN 0 7176 1115 9

Safe use of work equipment: Provision and Use of Work Equipment Regulations 1998. Approved Code of Practice and Guidance L22 HSE Books 1998 0 7176 1626 6

Safe work in confined spaces. Confined Spaces Regulations 1997. Approved Code of Practice L101 HSE Books 1997 ISBN 0 7176 1405 0

Safe working with flammable substances INDG227 HSE Books 1996 Single copies free, multiple copies in priced packs ISBN 0 7176 1154 X

Storing and handling ammonium nitrate INDG230 HSE Books 1996

Storage and handling of industrial nitrocellulose HSG135 HSE Books 1995 ISBN 0 7176 0694 5

Storage and handling of organic peroxides CS21 HSE Books 1998 ISBN 0 7176 2403 X

Storage and use of sodium chlorate and other similar strong oxidants CS3 HSE Books 1998 ISBN 0 7176 1500 6

The maintenance, examination and testing of local exhaust ventilation HSG54 HSE Books 1998 ISBN 0 7176 1485 9

The safe use of compressed gases in welding, flame cutting and allied processes HSG139 HSE Books 1997 ISBN 0 7176 0680 5

The spraying of flammable liquids HSG178 HSE Books 1998 ISBN 0 7176 1483 2

The storage of flammable liquids in containers HSG51 Revised HSE Books 1998 ISBN 0 7176 1471 9

The storage of flammable liquids in tanks HSG176 HSE Books 1998 ISBN 0 7176 1470 0

Use of LPG in cylinders Information Sheet CHIS5 HSE Books 1999

Use of LPG in small bulk tanks Information Sheet CHIS4 HSE Books 1999

References

Stationery Office (previously HMSO) publications

Department of the Environment and Welsh Office *The Building Regulations 1991: Approved Document B: Fire Safety* HMSO 1992 ISBN 0 11 752313 5

Estate Services Directorate. Northern Ireland *Firecode: Fire safety in residential care premises* HTM84 HMSO 1995 ISBN 0 337 07998 6

Health and Safety Commission/Home Office/Scottish Office *Guide to health, safety and welfare at pop concerts and similar events* HMSO 1993 ISBN 0 11 341072 7

Home Office *Safer communities: towards effective arson control. The report of the arson scoping study* Home Office 1999

Home Office/Scottish Home and Health Department *Guide to fire precautions in existing places of entertainment and like premises* HMSO 1994 ISBN 0 11 340907 9

Home Office/Scottish Office *Fire Precautions Act 1971. Guide to fire precautions in existing places of work that require a fire certificate. Factories, offices, shops and railway premises* HMSO 1993 ISBN 0 11 341079 4

Home Office/Scottish Office *Fire Precautions Act 1971. Guide to fire precautions in premises used as hotels and boarding houses which require a fire certificate* HMSO 1991 ISBN 0 11 341005 0

Home Office/Scottish Office/Fire Protection Association *Fire Precautions Act 1971. Fire safety management in hotels and boarding houses* HMSO 1991 ISBN 0 11 340980 X

NHS Estates *Firecode: Fire risk assessment in hospitals* HTM86 HMSO 1994 ISBN 0 11 321734 X

NHS Estates *Firecode: Fire safety in health care premises - general fire precautions* HTM83 HMSO 1994 ISBN 0 11 321725 0

Technical booklet on the Building Regulations (Northern Ireland) 1984. Fire safety HMSO 1984 ISBN 0 337 08 324 X

The Scottish Office *Technical Standards for compliance with the Building Standards (Scotland) Regulations 1990 as amended by the Building Standards (Scotland) Amendment Regulations 1993, the Building Standards (Scotland) Amendment Regulations 1994, the Building Standards (Scotland) Amendment Regulations 1996 and the Building Standards (Scotland) Amendment Regulations 1997* HMSO 1998 ISBN 0 11 495866 1

Arson Prevention Bureau publications

How to combat arson in schools Arson Prevention Bureau 1993

Prevention and control of arson in industrial and commercial premises Arson Prevention Bureau 1992

Prevention and control of arson in retail premises: A management guide Arson Prevention Bureau 1995

Other useful publications

A guide to good practice for the storage of aerosols in manufacturing, wholesaling warehouses and retail stores British Aerosol Manufacturers' Association 1989

Bulk storage of LPG at fixed installations: LPGA CoP No1, Part 1: Design, installation and operation of vessels located above ground LP Gas Association 1998

CIBSE Fire engineering guide E Chartered Institution of Building Services Engineers 1997 ISBN 0 900 953 78 0

Design approaches for smoke control in atrium buildings BR258 Building Research Establishment

Design principles for smoke ventilation in enclosed shopping centres BR186 Building Research Establishment

Fire protection measures in Scottish historic buildings Technical Advisory Note 11 Scottish Office ISBN 1 90 016841 3

Fire safety PB2281 Ministry of Agriculture, Fisheries and Food 1995

References

Inspection and testing Institution of Electrical Engineers Guidance Note No 3

Installation of sprinkler systems on historic buildings Technical Advisory Note 14 Scottish Office ISBN 1 90 016863 4

Protection against fire Institution of Electrical Engineers Guidance Note No 4

Safe practice for the storage of gases in transportable cylinders intended for industrial use Guidance Note GN2 British Compressed Gases Association 1988 ISBN 0260 4809

Scottish Education Department *A guide to fire safety in schools* Educational Building Note 18 Scottish Office 1982

Scottish Home and Health Department *Firecode in Scotland: Policy and principles* Scottish Office 1994 ISBN 0 7480 0975 2

Storage of full and empty LPG cylinders and cartridges LPGA CoP No 7 (Third edition) LP Gas Association 1998

The United Kingdom Working Party on Fire Safety in Historic Buildings *Heritage under fire: A guide to the protection of historic buildings* (Second edition) Fire Protection Association 1995 ISBN 0 902167 90 1

FURTHER INFORMATION

Here are some of the main organisations that can provide further advice.

Loss Prevention Council and **Fire Protection Association**
Melrose Avenue
Boreham Wood
Hertfordshire
WD6 2BJ
Tel: 0208 207 2345

British Fire Protection Systems Association
4th Floor
Neville House
55 Eden Street
Kingston-upon-Thames
Surrey
KT1 1BW
Tel: 0208 549 5855

British Approvals for Fire Equipment (BAFE)
4th Floor
Neville House
55 Eden Street
Kingston-upon-Thames
Surrey
KT1 1BW
Tel: 0208 541 1950

Building Research Establishment
Garston
Watford
Herts
WD2 7JR
Tel: 01923 664 000

Arson Prevention Bureau
51 Gresham Street,
London
EC2V 7HQ
Tel: 0207 216 7474

MENCAP
Royal Mencap
123 Golden Lane
London
EC1Y 0RT
Tel: 0207 454 0454

Further information

Royal National Institute for the Blind
224 Great Portland Street
London
W1N 6AA
Tel: 0207 388 1266

National Federation of the Blind of the United Kingdom
Old Surgery
215 Kirkgate
Wakefield
West Yorkshire
WF1 1JG
01924 291313

Royal National Institute for Deaf People
19-23 Featherstone Street
London
EC1Y 8SL
0207 296 8000

Chartered Institution of Building Services Engineers (CIBSE)
Delta House
222 Balham High Road
London
SW12 9BS
Tel: 0208 675 5211

Printed in the United Kingdom for The Stationery Office and the Health and Safety Executive
C1000   07/99